Wikipedia U

Tech.edu: a hopkins series on education and technology

Wikipedia U

Knowledge, Authority, and Liberal Education in the Digital Age

THOMAS LEITCH

Johns Hopkins University Press
Baltimore

Printed in the United States of America on acid-free paper
9 8 7 6 5 4 3 2 1

Johns Hopkins University Press
2715 North Charles Street
Baltimore, Maryland 21218-4363
www.press.jhu.edu

Library of Congress Cataloging-in-Publication Data

Leitch, Thomas M.
Wikipedia U : knowledge, authority, and liberal education in the digital age / Thomas Leitch.
pages cm. — (Tech.edu: A Hopkins Series on Education and Technology)
Includes bibliographical references and index.
ISBN 978-1-4214-1535-2 (hardcover : alk. paper) — ISBN 978-1-4214-1550-5 (electronic) — ISBN 1-4214-1535-6 (hardcover : alk. paper) — ISBN 1-4214-1550-X (electronic) 1. Education, Humanistic. 2. Research—Methodology—Data processing. 3. Wikipedia. I. Title.
LC1011.L44 2014
370.11'2—dc23 2014004984

A catalog record for this book is available from the British Library.

Johns Hopkins University Press uses environmentally friendly book materials, including recycled text paper that is composed of at least 30 percent post-consumer waste, whenever possible.

To Gardner Campbell

CONTENTS

ACKNOWLEDGMENTS

Michael Lonegro suggested the subject of this book. Laura Wimberley and Gardner Campbell provided invaluable ideas during the early months of its gestation. Fellows in two seminars I led in the Delaware Teachers Institute, "Media Literacy" in 2011 and "The Problems and Perils of Online Research" in 2012, explored some of its leading problems with me. Matt McAdam, Michele T. Callaghan, an extraordinarily helpful anonymous reviewer, and the enterprising and resourceful staff at Johns Hopkins University Press helped put it in its present form. Finally, I am forever indebted to numberless anonymous contributors, editors, and administrators of Wikipedia for providing me such a fertile and provocative field for research and so many inspiring examples of civilized disputation. It goes without saying, as it does in every Wikipedia page I have seen, that none of these generous and often unwitting collaborators is responsible for any errors of fact, omission, emphasis, or implication, which I hereby acknowledge as my own.

Wikipedia U

The Battle of the Books

DESPITE THE PROMINENCE of the word "Wikipedia" in this book's title, its true subject, indicated by its subtitle, is authority. Even if the title is not perfect in its emphasis, it is accurate enough, for the contemporary arena for debates about authority on which I focus is the intricate dance between the keepers of liberal education and the users and purveyors of online knowledge, embodied in the online encyclopedia Wikipedia. Throughout this book I will be generalizing frequently, and at times perhaps licentiously, about the role of liberal education in society, the ways online sources of knowledge are judged, and the question of who deserves to be called an authority. So it seems only fair to say as precisely as I can at the outset what I mean by each of these three concepts.

Dueling Experts: Wikipedia versus the Academy

I use *liberal education* to refer to the educational ideals and practices long associated with undergraduate colleges in the United States, which I will call "the academy." Undergraduate education has traditionally been devoted to the pursuit of what the noted theologian and education pioneer John Henry Newman called liberal knowledge: "knowledge which stands on its own pretensions, which is independent of sequel, expects no complement, refuses to be *informed* (as it is called) by any end, or absorbed into any art, in order duly to present itself to our contemplation."[1] A century later, Newman's ideal of the disinterested pursuit of knowledge rooted in a culture's collective heritage was reaffirmed for American education in a celebrated report compiled by Harvard University's Committee on the Objectives of a General Education in a Free Society. Although the report acknowledged the "obvious necessity for new and independent insights leading to change," it maintained the equal importance of "the impulse to rear students in a received idea of the good."[2] Shortly thereafter, Jacques Barzun, warning against the dangers of television, permissive education, and the anti-intellectualism of post–World War II America, quoted Newman's paean to "ratiocination" as best summarizing "the critical and co-ordinating functions of Intellect."[3]

Liberal education has for many years been the special province of liberal arts curricula. It is more closely associated with the social sciences than the hard sciences, more closely associated with the humanities than the social sciences, and most closely associated with areas like English and history and foreign languages that are not gateways to a single career in the way that chemistry or psychology can be. Liberal education has traditionally been associated not only with humanistic values, values stemming from universalistic assumptions about human nature, but, as the educational philosopher Harry S. Broudy points out, also with values as such: "If the sciences, physical and social, are the sources of warranted assertion as to matters of *physical* fact, the humanities claim to be the sources of warranted belief about *value* fact."[4]

Until recently the principal goal of liberal education—which Andrew Delbanco maintains springs from the ideals of religion and democracy[5]—has been what Richard P. Keeling and Richard R. Hersh describe as "not simply the active acquisition of knowledge, but also the active and increasingly expert use of that knowledge in critical thinking, problem solving, and coherent communication, as well as the personal, psychic, emotional, social, and civic learning of the student."[6] Louis Menand contends that "the academic's job in a free society is to serve the public culture by asking the questions the public does not want to ask, by investigating the subjects it cannot or will not investigate, by accommodating the voices it fails or refuses to accommodate."[7] Ideally, the result of the disinterested pursuit of knowledge in the service of an often unwilling public culture is the maturation of students through the mastery, integration, internalization, and responsible use of that knowledge, even if its fruits remain largely invisible. Hence, as Michael Bérubé maintains, "liberal education is fundamental to the future of democracy,"[8] for, in the words of former Harvard president Derek Bok, it "prepar[es] undergraduates to be democratic citizens."[9]

In our time, an increasing proportion of the knowledge that was long found in university libraries and other print archives is instead sought through online research—research that is conducted by means of computer-based digital resources, search engines, databases, archives, and other online tools. Most of the people who conduct this research do not happen to be academics. Travelers read reviews of hotels and restaurants before making their reservations. Do-it-yourselfers watch videos showing them how to install new showerheads and repair electrical appliances. Fans consult online sources for the latest theories about plotlines in *Game of Thrones*. I do not mean to marginalize these activities, which are far more ubiquitous than academic online research. I spend a good deal of time focusing on the relationship between liberal education and online research not because it is statistically the most frequent use of online resources but because it reveals most clearly the paradoxes of authority I wish to examine.

These paradoxes begin with the academy's love/hate relationship with the World Wide Web. The basis for the academy's love for the riches the digital age makes possible is everywhere. Online archiving has made digitized versions of priceless texts from the Dead Sea Scrolls to the Gutenberg Bible widely available to researchers and students. Computer databases and search engines make it possible to identify and locate scholarly resources with greater success, and in far less time, than earlier generations could have dreamed possible. The ability of university libraries to subscribe to online journals instead of procuring hard copies has freed sorely needed shelf space. The academy's constant hunger for new scholarly resources, protocols, and strategies—driven by the needs of both college teachers and the apprentices in their classrooms—has found its perfect complement in the Web. Even that lowliest of all computer technologies, word-processing software, has made it possible for scholars and students to write more quickly and accurately, revise more painlessly, and produce professional-looking copy more easily than ever before. For both students seeking material to use in fulfilling assignments and teachers engaged in original research, digital tools and resources have ushered in a golden age for the information that, as contemporary wisdom has it, wants to be free.

At the same time, the explosion of online research has provoked many conflicts. Scholars raised to trust books and journals often have serious reservations about online sources, even though, as Alvin Kernan reports, "More than half of my students were already telling me in the mid-1980s that what they saw on a computer screen held more truth for them than did a printed page."[10] What Kernan describes amounts to a contemporary version of the Battle of the Books Jonathan Swift dramatized in 1704 in a prologue to his satire *A Tale of a Tub*. Inspired by the debate between Bernard Le Bovier de Fontenelle and Sir William Temple over whether modern science and rationalism had eclipsed the wisdom of ancient Greece and Rome, Swift described a literal battle in the king's library between ancient and modern books that come to life bent on destroying each other. The same battle is restaged whenever advances in knowledge threaten the primacy of earlier authorities. For Kernan, the contemporary version of this battle is between digital publications that may never have been subject to prepublication review and codex books—books in traditional bound form—and hard copy journals threatened with what Sven Birkerts calls "a major sacrifice of authority."[11] This battle has been exacerbated by the appearance of online search engines in smartphones, whose size, portability, and association with a perceived deterioration of social and cultural norms make them anathema to many a seasoned researcher.

In addition, observers have often wondered whether the ready availability of information on the most recondite subjects has made students better or worse researchers, especially since it allows students to complete their homework more easily without developing the research skills that are the real point of the assignments. The power of search engines like Google to locate millions of Web pages discussing abortion, gay marriage, gun control, global warming, and intelligent design threatens to turn professional and especially amateur researchers into consumers of information. Like visitors to the exhaustive Library of Babel described in the prophetic short story by Jorge Luis Borges, they can be confident that every conceivable utterance on a given subject is archived somewhere on the Web, but the very proliferation of texts undermines the authority of any one of them. This crisis of authority has invaded the classroom, where students submit papers their teachers increasingly suspect have been purchased through the Web, downloaded from the Web, or cobbled together from Web sources. The explosion of unauthorized and open-access sites has raised troublesome questions about plagiarism, copyright, and intellectual property that may well inflect teachers' more general attitudes toward the Web's reliability.

Liberal education's skepticism about the online resources on which it has increasingly come to depend turns on the paradoxical nature of authority. Nowhere are authority's paradoxes sharper than in Wikipedia, the open-source online encyclopedia that has become the battleground of so many debates about the uses and abuses of online research in college classrooms. Many teachers categorically forbid their students to cite Wikipedia in their assignments, though this interdiction does not prevent students, or indeed the teachers themselves, from consulting Wikipedia without citing it. Wikipedia is the source everyone uses but no one is supposed to use or admits using. Its status as "the people's encyclopedia"—a phrase Sue Gardner, executive director of the Wikimedia Foundation, borrowed to describe it in a *Los Angeles Times* article marking its twelfth anniversary[12]—allows it to claim the kinds of authority associated with freedom and democracy but undermines its claims to the kinds of authority associated with professional competence and expert review. As new-media scholar Cathy N. Davidson has observed, however, "the discussion of peer review for collaborative knowledge-building sites such as *Wikipedia* throws into relief practices so widely accepted that we rarely question them anymore."[13]

Wikipedia is hardly unique in posing new paradoxes of authority and bringing older paradoxes to light. It is only the tip of the online iceberg, the most visible brand in the new wave of decentered authority the Web makes possible. Yet since its launch in January 2001, Wikipedia has been taken as both a bellwether and a scapegoat for debates about online research. It makes sense, therefore, to focus on Wikipedia in considering the paradoxical status of online authorities and the deeper paradoxes they reveal in the nature of authority itself. But in focusing on the relationship between Wikipedia and liberal education, I do not mean simply to use my own home base within the academy as a privileged lens through which to examine online research. In addition to using liberal education to frame the paradoxes of authority Wikipedia raises, I propose to reverse the process, using Wikipedia to reframe the paradoxes of authority implicit in liberal education as it plays out in the undergraduate classroom.

I have deliberately chosen Wikipedia and liberal education as codependent antagonists. Rather than taking the authority of either one for granted, I wish to use each one to discount the other. The philosopher and literary critic Kenneth Burke defines "discounting" as "making allowance for the fact that 'things are not as they seem.' . . . If a friend tells us something about ourselves, we discount the observation otherwise than we should if an enemy had made the same observation."[14] For Burke, every position taken on any subject, and ultimately every action, is based on motives and authorities that are situational, contingent, and subject to present discounting and future revision. Participants in these conversations begin by uncritically accepting the authority of a few chosen sources. As they discover a wider range of opinions, they often react against their earlier trust by a cynical "debunking" of all authority that seems more worldly but is equally indiscriminate.[15] Only by persistent engagement with a wide range of authorities can they develop the ability to discount authorities without simply dismissing them. Just as Wikipedia is a worthy representative of the strengths and weaknesses of online research in general, the college classroom and the habits and skills it seeks to inculcate are a logical choice to serve the role of the discounting critic whose skeptical embrace of online research's intrusion into the classroom is uniquely suited to reveal paradoxes in the authority of the dueling partners.

Liberal education has had a troubled relationship with online research for several reasons. Dedicated as it is to critical thinking and the production of new knowledge, it is clearly the redoubt that feels its mission most deeply threatened by the plagiarism, shoddy research, and lack of originality and oversight it fears will occur with those who look online for help in writing their papers and completing other assignments. The college teachers who condemn Wikipedia in particular in their classrooms, essays, and textbooks accurately perceive Wikipedia as incarnating a bottom-up model that directly challenges the top-down model of authority that has long been the basis of graduate and professional education. This top-down model assumes that education, as the word's etymology suggests, is a process whereby professors free students' minds by leading them away from flawed, erroneous, or incomplete knowledge to a deeper and truer knowledge to which their professional training and expertise has given the professors access.

In addition to providing a direct challenge to the research model prevalent in the academy since its inception in the Middle Ages, and increasingly widespread in American research universities since the later nineteenth century, Wikipedia offers an unexcelled laboratory for examining and comparing different models of authority. Its relatively loose centralized editorial control allows its millions of entries to exemplify a wide range of models that often contradict each other and illuminate similar contradictions in the range of models of authority liberal education has adopted in practice. Just as liberal education provides the ideal forum for discounting online research, Wikipedia provides an ideal instrument for discounting contemporary liberal education, particularly for probing the well-publicized conflicts between teachers and students, or more broadly between educators and the people who pay their salaries.

Even before Wikipedia's founding at the turn of the twenty-first century, the widening economic division between American workers with and without college degrees already made such a degree seem increasingly imperative, allowing colleges to charge more and more for this credential, if not for the skills or knowledge it betokens. At the same time, it made colleges compete more aggressively with each other in terms of the amenities they offer or the economic edge their degrees can be expected to provide. According to Richard Arum and Josipa Roksa, liberal arts colleges, while they officially espouse "teaching students to think critically and communicate effectively . . . as the principal goals of higher education," confer passing grades and degrees on unmotivated students who have little desire to master new ways of thinking, because "existing organizational cultures and practices too often do not prioritize undergraduate learning."[16] Arum and Roksa quote education researcher George D. Kuh on the "disengagement contract" teachers offer their students: "'I'll leave you alone if you'll leave me alone.' That is, I won't make you work too hard (read a lot, write a lot) so that I won't have to grade as many papers or explain why you are not performing well."[17]

Even when this contract is not in effect, teachers and students are often at odds with each other, for students who enter college expecting to emerge with a credential that will substantially increase their lifelong earning power will adopt sharply different attitudes from those of their teachers toward assignments that are designed to foster skills and habits that seemingly have no clear material value. They will resist jumping through hoops to train for obsolete careers. They will be particularly uninterested in the pursuit of knowledge for its own sake or for the sake of the intellectual exercise of the pursuit. And they will chafe against any restrictions on their use of information.

A Brief Anatomy of Authority

Perhaps because it is so much more difficult to define than liberal education or online research, authority is a concept whose definition is surprisingly little discussed or debated in the contexts in which it is most often invoked. Disputes among nations, political campaigns, science, and organized religion all involve assumptions about authority. Most of the parties involved, however, are typically more interested in asserting their authority and defending it against all opponents, especially previous authorities they seek to dethrone, than in encouraging partisans to adopt more critical attitudes toward authority.

This is dispiriting in view of the many fields to which authority is clearly central. Military and political histories, along with histories of language and science, are almost by definition shaped by debates among conflicting authorities: Athens versus Sparta, Ptolemy versus Galileo, evolution versus creationism. Literature abounds in conflicts among authorities. Most of Shakespeare's history plays, along with *Hamlet, Macbeth,* and *Julius Caesar,* turn on questions of kingly authority: Who has the right to rule? Can a king by bad behavior forfeit this right? What gives someone the right to depose a king and rule in his place? Legal history and legal studies generally are organized around the successive resolutions of competing claims to authority. Even religion, which tends to ascribe categorical authority to its founding texts, is surprisingly rich in debates over authority. Moses and Aaron bolster their initial appeal to Pharaoh to free the Israelites to leave Egypt by turning Aaron's rod into a serpent. Although Pharaoh's sorcerers can also turn their rods into serpents, Aaron's serpent swallows all the others.[18] Later, the prophet Elijah converts the followers of Baal to belief in God by arranging a head-to-head competition in which God works a series of miracles that Baal cannot equal.[19] The Gospels are full of debates pitting the authority of Jesus against that of Jewish or Roman law, debates that reach a climax when Jesus is brought before Pontius Pilate.

Authority is not the same thing as power, since lynch mobs, for example, can exercise power without authority. Nor is it the same thing as legitimate grounds for the exercise of power, for rulers can be deposed even if they remain in possession of such legitimate grounds. Authority is best defined as the *recognition* of legitimate grounds for the exercise of power. As Max Weber, the preeminent theorist of authority in the twentieth century, observes: "Experience shows that in no instance does domination voluntarily limit itself to the appeal to material or affectual or ideal motives as a basis for its continuance. In addition every such system attempts to establish and to cultivate a belief in its legitimacy."[20] Although authority is treated as a problematic notion in philosophers from Plato to Descartes, it does not emerge as a practical problem until the Reformation. When the need for political authority can no longer be taken for granted, as it was in ancient Greece and Rome, or referred to the divine right of kings to rule, as it was in the Middle Ages, the question of what justifies or legitimizes authority figures—what gives authorities their authority—becomes central and urgent to all discussions of power, leadership, and belief.

Philosophical discussions of authority customarily distinguish *epistemic authority*, the authority offered for some particular knowledge or belief, from *moral authority*, the right to govern the behavior of others. Moral authority focuses on perceptions of the legitimacy behind the people and institutions who exercise power, especially political power; epistemic authority, focusing on the search for authorities who might legitimize one's beliefs and actions, seeks grounds for understanding why we believe what we believe and not something else and why we change our minds when we do. Weber distinguishes "three pure types of legitimate domination": *rational authority*, "resting on a belief in the legality of enacted rules"; *traditional authority*, "resting on an established belief in the sanctity of immemorial traditions"; and *charismatic authority*, "resting on devotion to the exceptional sanctity, heroism or exemplary character of an individual person."[21] It is clear from this analysis that all modes of moral authority are social, deriving their force and meaning from the specific communities who create them by recognizing them.

Weber's three modes of moral authority are equally relevant to the epistemic authority that is central to the understanding of both Wikipedia and liberal education. Though epistemic authority is not the same thing as power, it establishes a hierarchy of power within a given community in which authorities stand higher than nonauthorities. Epistemic authority, like political authority, is unequal and asymmetrical, since granting authority to someone else involves ceding one's own claim to that authority. Both kinds of authority must be conferred from outside. Dictators and professional experts can seek authority, but they cannot confer it on themselves, as Julius Caesar's status as emperor, for example, required acceptance by the Roman Senate and people. Even more clearly than political authority, epistemic authority is hypothetical. It must be granted or assumed or tacitly stipulated rather than seized.

Contemporary philosophers of authority continue to wrestle with the problems it raises. Richard T. De George hints at a crucial insight when he observes that "X is not usually an authority for those who certify him, since they must generally have superior knowledge in order for their certification to be acceptable to others."[22] Epistemic authority requires a distance between a given authority and those who accept this authority. As E. D. Watt puts it: "In some cases authority cannot function if there has been a complete explanation of the matter in hand: a person who understands a pronouncement completely, with all the reasons for it, can no longer accept it on authority (though he may still accept it), and Dr. Johnson's dictionary cannot be consulted as an authority by Dr. Johnson."[23] Epistemic authority depends on distance, absence, and deferral. It is defined by the gaps, limits, and pockets of ignorance that encourage people to appeal to it.

Just as moral authority is not the same as power, epistemic authority is not the same as knowledge. It is more properly defined as the recognition of relevant knowledge or legitimate power by others. As phrases like "valid authority" or "just and rightful authority" indicate, authority is not always viewed as just, rightful, or valid; if it were, it would not need the help of these adjectives to substantiate its claims. Authority—whether rational, traditional, or charismatic—is more closely aligned with trust than with power, knowledge, or rights.

The claims of all authorities are subject to challenge by competing authorities. Disconfirming a given authority is a paradoxical process, since an authority can be disconfirmed only by reference to another, higher authority—for example, "the facts" as conveyed by one's own senses or a more trusted source. Confirming a given authority is equally paradoxical, not only because it involves invoking another authority but, because once confirmed by appeals to reason or evidence, an authority no longer requires the trust that defined it as an authority. Authority is evidently a developmental relation whose power and relevance depend on unequal parties' not yet having achieved complete understanding, communion, or unity.

Even if it is difficult to make firm distinctions among the concepts of epistemic authority, influence, knowledge, and power, philosophers have made useful distinctions between different modes of epistemic authority. Richard Foley distinguishes *fundamental authority*, based on the disposition "to be influenced by others even when we have no special information indicating that they are reliable," from *derivative authority*, which springs from the audience's belief that a given source's "information, abilities, or circumstances put [him] in an especially good position" to earn trust.[24] Dennis H. Wrong goes so far along these lines as to distinguish authority as "the untested acceptance of another's judgment" and persuasion as "the *tested* acceptance of another's judgment."[25]

The most obvious way a given authority can establish its superior, derivative authority without submitting its judgment to testing is through a claim of expertise. Indeed, claims to expertise are so ubiquitous in liberal education that it is surprising to see these claims relatively marginalized in Weber's tripartite anatomy of authority. But in fact Weber's analysis reveals just how narrow these claims are. Weber considers expertise a desideratum in the legal bureaucracies whose claim to authority is rational. The modern university, in his view, is just such a bureaucracy, providing "the kind of 'education' which is bred by the system of specialized examinations or tests of expertise (*Fachprüfungswesen*) increasingly indispensable for modern bureaucracies." Weber acknowledges the "ambivalent attitude" democracies adopt toward such potentially elitist examinations and notes that "social prestige based upon the advantage of schooling and education is by no means specific to bureaucracy." Although he notes that "behind all the present discussions of the basic questions of the educational system there lurks decisively the struggle of the 'specialist' type of man against the older type of the 'cultivated man,'" he sees this struggle as "conditioned by the irresistibly expanding bureaucratization of all public and private relations of authority and by the ever-increasing importance of experts and specialised knowledge." Weber assumes that the formation of the "cultivated man" depends on "the educational ideal stamped by the structure of domination and the conditions of membership in the ruling stratum of the society in question." In this model, the "*cultural quality*" of the cultivated man is a "plus" in the same way that "expert knowledge" is a "plus,"[26] something education has added to students.

Writing from within a culture comfortable with its drive toward ever-increasing expertise, Weber does not consider the foundational paradoxes of the contemporary academy, which seeks to establish itself as both dispensing and transcending expertise. Professors teach students to establish their own authority through their success in asking better questions of established authority. At the same time these professors are dispensing the consensual wisdom of their disciplines in class, they are working to challenge that conventional wisdom or expand its boundaries in their own research. The resulting paradoxes are beyond Weber's scope. But his analysis of democracies' ambivalence toward educational expertise prophesies America's ambivalence toward the academy along with the academy's own ambivalence toward the educational goals of professional expertise and general acculturation, which advance very different claims to authority.

Even academics less critical of their own claims to authority commonly make distinctions among the different kinds of authority in their fields—for example, between primary and secondary sources, those like eyewitness accounts or laboratory reports whose authority is direct and those whose authority depends on the other authorities they summarize or cite. Teachers forbidding their students from citing Wikipedia often describe it as a good place to begin research, implicitly distinguishing its authority from the final authority of sources that are presumably a good place to end research. These teachers evidently think of Wikipedia as comparable to the spotting scope attached to the top of an astronomical telescope, which uses low magnification to locate planets and stars more easily in the night sky. Once people have located these celestial bodies with the scope, they can study them more closely by using the high-magnification telescope. More generally, poststructuralist philosophers and analysts have distinguished between the authority of *logos,* the spoken word associated with a teacher like Jesus, whom John's Gospel calls the Word of God, and *lexis,* the written word associated with an absent teacher whose authority relies on the mediation of a text designed to substitute for the teacher's presence. Both Deuteronomy, the fifth book of the Pentateuch, and the Acts of the Apostles, the fifth book of the New Testament, prefigure the modern transition from *logos* to *lexis* as the basis of authority. When God ceases to speak to his people directly, the divine prophets or followers who speak on his behalf often feel obliged to bolster their authority—or, in Wrong's terms, to substitute persuasion for authority—by providing reasons and justifications for their directives that God does not feel obliged to provide.

Authority may be hard to define, but it is easy to recognize its multiple modes as they are revealed by conflicts: in wars, in courtrooms, in political clashes, in crises of authority, and in the kinds of questions often raised by liberal education. Whenever authorities compete with one another, a set of fundamental questions arises: How should we choose among conflicting authorities? What gives each authority its authority? How can we check on the authority of different authorities? These questions are especially urgent in online research because of a flattening effect the Web confers on all entries. Just as every book in a library looks equally trustworthy as long as they all remain on the shelf, every Web page has the potential to look as trustworthy as every other. Online sources that are filled with logical flaws, fallacies, and factual errors can look trustworthy if their architecture and visual design look professional. In addition, it is often difficult to check the authority claims of online sources, because often they do not provide references or citations that would establish clear grounds for those claims. More fundamentally, the Web's claims to authority are, in Richard Foley's terms, primarily fundamental rather than derivative. They depend on what Evan Selinger and Robert P. Crease call "the democratic and antielitist urge to accord equality to all opinions."[27] This urge is often exaggerated by people who go online looking for answers to specific questions but stop looking after reading the first answer they find.

Though the academy is highly sensitive to problems of authority in online research, it is much less inclined to examine the problems implicit in its own claims to authority. Medieval universities rooted their authority in divine scripture; modern universities root their own in the scholarly and professional expertise represented by their faculties. The kinds of authority most explicitly valued in liberal education, which depend on sources and citations, would seem to favor *lexis* rather than *logos*. But this model is complicated by several factors. The authority commonly accorded individual scholars depends not only on the sources they cite but the frequency with which they are cited themselves, which reflects the recognition and esteem they are accorded by peers and other authorities. Unlike the Web, which merely offers a home to competing authorities, the academy actively encourages debates among them, not so much in order to settle those debates once and for all as to use them to drive the disciplines that depend on constant intellectual ferment. Series like Blackwell's *Great Debates in Philosophy* and Prentice-Hall's *Twentieth-Century Interpretations,* both aimed at college students, showcase competing claims in the hope of keeping them alive rather than settling them for good.

Despite the academy's nominal dedication to *lexis,* students often seek the authority of a reassuring *logos.* Accommodating classroom teachers, for better or worse, are often happy to oblige them by supplementing or replacing the expertise acknowledged by their professional peers with the kind of charismatic authority defined by Weber and duly recorded by websites like RateMyProfessors.com. Although such teachers may fail to teach their students to question the teachers' own authority, one of the primary values of a liberal education is the practice it gives students in asking intelligently critical questions about authority without simply defying or rejecting it—assuming of course that they can learn to question the authority of liberal education itself, which, like the Army and the Church, is more sensitive to the limits of other authorities than its own.

Throughout this book, my analysis of the contemporary Battle of the Books focuses on examining other parties' claims to authority rather than defending my own, except of course by implication. My view of epistemic authority assumes the public acceptance of a claim to moral or ethical or intellectual legitimacy. In Weber's terms, I believe that epistemic authority is essentially rational rather than traditional or charismatic, though I acknowledge the force of both these other modes of authority, especially in liberal education. And I follow Richard Foley in grounding the operation of authority in specific social institutions, from the American electorate to the followers of a blog to a single teacher's classroom. The kinds of authority in which I am most interested—and the claims to authority that I examine—are not transcendent or timeless but temporal, rationally grounded, historically situated, and subject to change. This definition may seem narrow, but it is shared by both the academy and the Web.

The Use of Crises in Authority

According to Barack Obama's former chief of staff Rahm Emanuel, you should never let a crisis go to waste. Crises in authority are invaluable because they reveal the paradoxical nature of different definitions of authority, and of authority itself, with exceptional power and clarity. The crisis of authority online research poses for liberal education is only the latest of many such crises. Observers have most frequently compared the digital revolution to the Gutenberg revolution that replaced the authority of the text handwritten by someone who presumably had direct control over both its contents and its composition with the more dubious authority of a text produced by a technology that introduced untold new possibilities for error. As late as the nineteenth century, two rabbis of Slovita, Reb Shmuel Abba and his brother Reb Pinhas Shapira, sought to indemnify the publishing concern they had inherited from their father, a firm restricted to printing sacred texts, from error by purchasing a new printing press, "cart[ing] the entire press to the *mikveh* in Slovita and dipp[ing] each part of the press in the *mikveh*."[28]

The pattern of using the authority of the old technology, in this case the ritual bath designed to purify Orthodox women once a month, to certify the new technology, however inappropriately, is so common that it raises the question whether there is ever in fact not a crisis of authority. Although historian of science Thomas Kuhn and literary theorist Stanley Fish both define paradigm shifts in contrast with periods of stability in what Kuhn calls "scientific communit[ies]" and Fish "interpretive communit[ies],"[29] discursive arenas like courts of law and scholarly quarterlies so inveterately weigh the competing claims of different authorities that for them, such crises are literally the order of the day or the quarter. One way to resolve this apparently endless sense of crisis in legal and scholarly debate is to normalize it and minimize its sense of conflict, as in Isaac Newton's famous dictum that, "if I have seen farther, it is by standing on the shoulders of giants." Indeed, this sentiment had itself, as Robert K. Merton has shown, both a long history before Newton and a contentious edge, "a motivated hostility toward a forerunner,"[30] that its Olympian modesty imperfectly conceals.

Bob Young of Red Hat Software notes that the free software movement represented by open-source operating platforms like Red Hat's Linux has given new relevance to Newton's aphorism:

> "In the western scientific tradition we stand on the shoulders of giants," says Young, echoing both [Linus] Torvalds and Sir Isaac Newton before him. "In business, this translates into not having to reinvent wheels as we go along. . . . If you need a graphic tool set, you don't have to write your own graphic library. Just download GTK [Gimp Tool Kit]. Suddenly you have the ability to reuse the best of what went before. And suddenly your focus as an application vendor is less on software management and more on writing the applications specific to your customer's needs."[31]

In both Merton's and Young's readings, Newton's aphorism is intended not so much to deny or resolve conflicts in authority as to manage them for the benefit of future scientists or present-day customers. Whenever crises in authority arise—whether in the course of momentous scientific revolutions or in the daily business of courts weighing the claims of one party and series of legal precedents against another opposing party and series—partisans of the new and old orders alike seek to manage them, typically for the sake of advantageously ordering them. At the same time, these crises offer observers who are invested in authority and interested in thinking about it new opportunities to press questions that are not clearly slanted toward either side. Recent debates between gun-control advocates and defenders of the Second Amendment, for example, offer opportunities to consider potent general questions behind the frequent posturing on both sides. What gives utterances and institutions like the Bill of Rights their authority and why? Why do we look for authority and value it when we recognize it? Why is it good, and what is it good for? Given the authoritative nature of authority, why do modes and fashions of authority change instead of remaining constant? Why do crises in authority appear when they do? Why are some challenges to authority recognized as more momentous or far-reaching than others, and who decides when and how the nature of authority has shifted? And given the mutable nature of authority, is authority ever truly stable, or merely stable-seeming to partisans who are invested in ignoring or repressing challenges to it?

The contemporary Battle of the Books between the academy and the print authorities with which it is associated and the online universe, which conceives authority in very different terms, puts a new spin on these evergreen questions and gives them a new urgency. But they cannot be resolved simply by taking sides with the Ancients or Moderns of the twenty-first century. Exploring problems of authority requires us to discount our modern Battle of the Books by looking beneath it or rising above it, beginning with a critical examination of stories that have been told to account for its origins.

CHAPTER ONE

Origin Stories

FUNDAMENTAL QUESTIONS ABOUT AUTHORITY are given sharper definition by the origin stories that purport to explain how new authorities came to arise and assume sufficient strength to challenge their predecessors. The case of Wikipedia is especially illuminating, because it has generated four distinct genetic narratives of Wikipedia's ancestors and birth. Each of them proposes a history designed to manage questions of authority that Wikipedia raises by setting the world of Wikipedia against the world before it existed. Each of them has an agenda designed to buttress or question the authority of Wikipedia. Each of them willy-nilly reveals paradoxes of authority that the others leave unexamined. Each of them is illuminating about which genetic analogies it deems most appropriate to explain the rise of Wikipedia, what kinds of authority it claims for online research in general and Wikipedia in particular, and what assumptions it makes about the nature of authority. And all of them are valuable for the more oblique light they throw on the nature of authority in contemporary liberal education.

The First Narrative: Wikipedia Presents Wikipedia

The first of these narratives, Wikipedia's institutional account of its own status most familiar from the Wikipedia portal's self-branding, is an origin story in only a limited sense. The URL www.wikipedia.org takes users to an English-language page dominated by a graphic of an incomplete spherical puzzle every piece of which is marked by a letter in a different alphabet. This graphic is circled by hotlinks to ten leading Wikipedia portals in different languages—in English, French, German, Italian, Polish, Spanish, Russian, Japanese, Portuguese, and Chinese—each labeled "the Free Encyclopedia" in its own language, each indicating the number of articles currently available in that language, with English-language entries currently over four and a half million. Beneath this array is a search engine that allows users to choose any of fifty-three languages. Further down is a series of hotlinks to the Wikipedia portals in each of the nine languages in which more than a million articles have been posted, then to the portals in each of the forty-three additional languages in which more than a hundred thousand articles have been posted, then to the portals in each of the seventy-three languages in which between ten thousand and a hundred thousand articles have been posted, then to the portals in each of the one hundred one languages in which between a thousand and ten thousand articles have been posted, then to the portals in each of the fifty-one languages in which between a hundred and a thousand articles have been posted, and then to a list of "other languages" that sends users to a page that lists all 287 languages in which Wikipedia portals have been opened, including nine to which fewer than one hundred articles have been posted and a tenth, the Herero language of Namibia, for which the only listing is an English-language page about the Herero language itself. The very bottom of the Wikipedia portal includes hotlinks to Wiktionary, Wikibooks, Wikisource, Commons, Wikinews, Wikidata, Wikiversity, MediaWiki, Wikiquote, Wikispecies, Wikivoyage, and Meta-Wiki, and a button that takes users to the home page of the Wikimedia Foundation.

The puzzle-globe graphic in this portal suggests the global, cooperative nature of Wikipedia and its status as a work-in-progress. The page's lexical elements emphasize the project's global reach but add an element of competition among different languages as their users presumably seek to rise higher in the list by adding more articles.[1] The links at the bottom of the page promote the Wikipedia franchise by highlighting the foundation's other related activities. The sense of authority these items communicate is not rooted in reliability and stability but in the promise of ceaseless growth and expansion. At first blush, the narrative it generates would seem to be an antiorigin story because it looks forward rather than back. The paradoxical promise that an online encyclopedia can be authoritative because of its orientation toward the future rather than the past is underlined by the portal's prominent reference to "the Free Encyclopedia." This is virtually the only text on the page that does not serve a utilitarian directive function and seeks to distinguish Wikipedia from other encyclopedias. At the same time that it looks forward, however, the portal takes care to root its claims to authority in the more traditionalist claims of those earlier encyclopedias. Each listing of languages in which Wikipedia operates is surmounted by a graphic of a bookshelf, longest in the case of languages with more than a hundred thousand entries, shortest in the case of languages with between a hundred and a thousand. Using the length of bookshelves to indicate how extensive the entries in different languages are allows Wikipedia both to borrow and to disavow the iconography of the print resources with which it competes.

This first narrative of Wikipedia's origins implicitly contrasts it with print encyclopedias. The contrast is made explicit in the afterword of Wikipedia administrator Andrew Lih's volume *The Wikipedia Revolution*, which argues that print encyclopedias have three significant disadvantages: their cost means that access to them, or at least ownership of them, is "limited by income"; it would take "too much physical space and expense" to prepare a "truly comprehensive printed encyclopedia"; and it is difficult to keep "a printed encyclopedia of any size accurate and up-to-date."[2] Within five years of its launch in January 2001, Wikipedia had consciously begun a drive to compete successfully with print encyclopedias in these three areas. Since all users of Wikipedia could correct any errors they discovered and every article could be updated in real time, Wikipedia carried the potential to be much more timely than any print encyclopedia. The many entries Wikipedia devoted to topics neglected by print encyclopedias—especially topics concerning contemporary politics, popular culture, and other subjects more timely than universal—made Wikipedia, for better or worse, far more comprehensive than its print counterpart. And this staggering amount of information was available to any user with access to an uncensored computer terminal for free.

This first narrative, then, is capped by Wikipedia's success in beating print encyclopedias at their own game by presenting itself frankly as a work-in-progress. The ability of its worldwide network of users to add, update, and correct material in millions of entries in hundreds of different languages—information that Wikipedia allows to be freely shared among all users—does not compromise but actually improves the vast trove of information it archives. By embracing its status as a work-in-progress, Wikipedia grafts one central value of liberal education, the need for constant critical attention and review, onto another equally central value, the need to preserve and respect the authority of the past. In doing so, it seeks to eclipse the authority of print encyclopedias whose repository of knowledge, updated only every ten years or so, represents the second of these values far better than the first. Wikipedia's relationship with print encyclopedias will play an equally prominent but quite different role in the fourth narrative.

The Second Narrative: Wikipedia and Wikis

The second narrative traces Wikipedia's development not from earlier print encyclopedias but from the wiki, a collaborative online format for documents-in-progress first developed in 1995 by software developer Ward Cunningham, who thought of it as "a moderated list where anyone can be moderator."[3] In this narrative, Wikipedia's self-branding as "the Free Encyclopedia" cuts much deeper than its availability at no monetary cost. In his foreword to Lih's volume, Wikipedia cofounder Jimmy Wales emphasizes that this label is meant to emphasize "free as in speech, not free as in beer. . . . When we talk about Wikipedia being a free encyclopedia, what we're really talking about is not the price it takes to access it, but rather the freedom that you have to take it and adapt it and use it however you like."[4] Wales's assertion seems to respond to legal activist Lawrence Lessig's warnings about the antilibertarian drift and dangers of the "code" that constitutes the "law" of cyberspace.[5] Wales's list of "four freedoms" Wikipedia gives its users—"You get the freedom to copy our work. You can modify it. You can redistribute it. And you can redistribute modified versions"[6]—links the ability of every user to stand on the shoulders of giants to a specifically American narrative of democracy and freedom. This narrative, linking the American ideal of democracy to the online ideals of universal access and user-friendliness, is given dramatic shape by the subtitle of Lih's volume, *How a Bunch of Nobodies Created the World's Greatest Encyclopedia,* which invokes a quintessentially American parable of the underdog triumphing over impossible odds. Wales picks up this pattern when he cites Wikipedia's global appeal, the vast number of articles it makes available in an impressive array of languages, and its prodigious growth curve, noting that it is "several times larger than Britannica and Encarta combined" and exultantly concluding: "We see more people, or more people see us, than the *New York Times*; we see more people than the *LA Times,* the *Wall Street Journal,* MSNBC.com, and the *Chicago Tribune.* The really cool thing is, we see more unique visitors in a single day than all these sites combined."[7] What makes this really cool, and not just really interesting, is Wikipedia's status as the upstart triumphant.

This narrative of Wikipedia's genesis reads its moral authority, its status as David versus Goliath, as communal power. The ability of Wikipedia to create a free and democratic community is reflected in the invitation Wales routinely extends in speeches and interviews: "Imagine a world in which every single person is given free access to the sum of all human knowledge. That's what we're doing."[8] Wikipedia seeks both to express and to transcend the authority of any individual contributors through what Joseph

Michael Reagle Jr. calls "a *good faith* collaborative culture."[9] By collecting a critical and growing mass of wikis whose ultimate goal is to share all the information its users have or want, Wikipedia establishes itself as the wiki that is not content to be simply a wiki, a collection of wikis whose goal just happens to comprise the sum of human knowledge. Its contributors are both more and less anonymous than are the contributors to print encyclopedias, whose names are often readily available at the end of the articles they have written but ignored by readers who take the encyclopedia, not its contributors, as the ultimate source of authority. The Wikipedia community is one in which all members can serve as contributors and editors unless they run afoul of rules designed to preserve equality among contributors, rules that have inescapably made some contributors more powerful than others.

Wales's extravagant claim flies in the face of liberal education's painstaking distinctions among information, knowledge, and wisdom. One of the most common critiques of online research is that Web boosters like Wales conflate information, which the Web offers in unprecedented abundance, with knowledge and wisdom, which it does not—or at least offers so partially and selectively that users, like Borges's patrons at the Library at Babel, would already need to possess a good deal of knowledge and wisdom to discriminate between sites that were more and less wise and knowledgeable. Another critique concerns the distinction, which is once again muddled by Wales's conflation of information and knowledge, between two senses of Internet browsing. Traditionally, to browse means to look around with a view toward taking in the widest possible range of experience, like a window shopper or a library patron scanning titles in the stacks. Internet browsers like Safari and Firefox and general search engines like Google and Bing are designed to promote this goal of making the widest possible array of pages available to viewers. In practice, however, most Web searches are conducted by people seeking answers to specific questions: When did John Lennon die? How far is it from Winnipeg to Montreal? What are the chances of rain tomorrow afternoon in San Francisco? How is the euro doing against the dollar? These people are not browsing; they are seeking information, not experience, knowledge, or wisdom. Google, the search engine designed to steer users to the pages most likely to contain the information they need to answer specific questions, features a hotlink labeled "I'm Feeling Lucky" that takes online window shoppers searching for a particular term or string to a site on which that text appears. The appeal to luck might seem to suggest that the selection of the site will be random, but in fact the hotlink is designed to take visitors to the top search result—very often, the Wikipedia page on the subject—every time. Luck has nothing to do with it.

Wales's conflation of information with knowledge rejects the Web in general and Wikipedia in particular as a cyberspace where numberless solitary individuals seek answers to individual unrelated questions in favor of an online community that transcends parochial and political boundaries because its mission is to "contribute knowledge to the world." For Wales, therefore, "Wikipedia isn't a technological innovation at all; it's a social innovation."[10] Like Ward Cunningham, Wales assumes that wiki contributors create intellectual communities whose authority is greater than that of any given contributor. By continually reconstituting authority on an ad hoc basis, Wikipedia replaces the hierarchical model of institutional authority typical of encyclopedias, libraries, and university administrations with a decentralized model that returns authority to the grassroots communities it creates or restores. The authority of Wikipedia exceeds that of other wikis not only because of its great number of entries but, even more important, because of its great number user-editors. These contributors are statistically more and more likely to correct for the intervention of vandals and cranks, in accord with a peculiarly American faith in the marketplace of ideas, in which, contrary to Gresham's law predicting that bad money will flood out good, good ideas inevitably drive out bad.

As Wales and Lih acknowledge, the radically democratic impulse Wikipedia shares with other wikis is tempered in practice by a series of editorial protocols and controls. Although every user of Wikipedia is equal, some are more equal than others. Instead of attempting to resolve or dissolve this paradox, Wales attempts to rise above it by his emphasis on Wikipedia as a knowledge community whose collective authority transcends battles for authority among individual users, even though the outcomes of these battles determine Wikipedia's collective voice. Wales waxes utopian in his view of Wikipedia as a template for other such communities. Having argued that "the software does not determine the rules of Wikipedia. . . . There's very, very little in the software that serves as rule enforcement. It's all about dialogue, it's all about conversation, it's all about humans making decisions," he concludes: "Let's take these ideals of Wikipedia and bring them out to lots and lots of people in lots and lots of areas far beyond simply encyclopedias. I think the genuine communities, like Wikipedia, will be built on love and respect. But it's really important . . . to remember that Wikipedia is not about technology, it's about people. It's about leaving things open-ended, it's about trusting people, it's about encouraging people to do good."[11]

This apotheosis and its implicit exhortation recall nothing so much as the discourse of authority in democratic elections. No leader is perfect, but leaders elected by a majority of voting citizens best express their collective will, even to their collective contradictions, as long as the citizens inform themselves about the issues and take their responsibilities as voters seriously. Their informed participation gives their leaders the authority to claim mandates on behalf of the voters who elected them ("the people have spoken"), even when they make decisions that may affront the wishes of those voters. The informed electoral participation of each citizen is the presumptive link on which democratically elected leaders' authority depends. In the same way, the authority of liberal arts colleges depends not only on the received wisdom they make available for their students' consumption, deliberation, and critical revision but also on the track record of their alumni, on their success in living and sharing the values they have learned. The value system Wikipedia shares with other wikis is not a challenge to liberal education but its logical extension into cyberspace, because Wikipedia does not resolve these paradoxes of authority in liberal education but faithfully replicates them.

The Third Narrative: Wikipedia and Personal Computers

A third narrative of Wikipedia's origins is implicit in attacks on and defenses of what has come to be called Web 2.0. This narrative has its roots in the dawn of the computer, or the analytical engine, as Charles Babbage called it in 1837 when he envisioned a machine that could be programmed by means of punched cards to calculate polynomial functions, square roots, and multiples of *pi*. Babbage never succeeded in constructing an analytical engine. The first computers that were constructed a century later—although they incorporated the kinds of algorithms British mathematician Alan Turing described that could reduce complex operations to a series of discrete logical steps—were similarly restricted to mathematical computations. Even the Colossus, used by the British to decipher code messages encrypted by the German military during World War II, did so by analyzing Boolean functions in different encrypted transmissions. Numerical calculators had become fixtures on the desks of mathematicians and accountants by the 1960s. But it was not until the 1970s—after steady increases in processing power, the development of programming languages like FORTRAN, COBOL, and BASIC, and the replacement of the extensive circuitry early computers required by microprocessors—that computers like the Wang 2200, the Xerox Alto, and the IBM 5100 were widely marketed to businesses. The rapid multiplication of microprocessors' ability to store data brought down the price of computing enough for the first successful personal computers, the Commodore PET, the Apple II, and the Radio Shack

TRS-80, to be marketed in 1977. The first widely available word processing program, WordStar, followed in 1979.

By the early 1980s, families fortunate enough to own personal computers could use them to type and revise documents, prepare budgets and other spreadsheets, and play games like Pong or, if they had a dedicated gaming console, Space Invaders and Pac-Man. The development of the computer from Babbage's original design for an analytical engine to the rise of the personal computer is marked by a turn from regarding computer users as programmers and experts, professional colleagues of computer designers, to regarding them as consumers who could master the common functions of computers without knowing anything about programming, debugging, or troubleshooting. Even when word processing software like WordPerfect and Microsoft Word departed from the one-size-fits-all platform and formatting of WordStar and allowed their users increasing ability to customize menus and functions to suit themselves, computer literacy was understood as the ability to run software programs, not design, alter, or repair them. A decade later came the advent of open-source platforms like Linux, whose devotees prided themselves on their ability to tinker under the hood. In spite of this, the personal computer became a black box to most users. The widening gap between designers of computer hardware and software and the consumers for whom that hardware and software were intended was widely regarded as the necessary price to insinuate the computer into every home that could afford it.

This shift might be described by analogy to two distinct models of instruction in college classrooms: the doctoral seminar and the undergraduate lecture. In the days before home computers, computer users were the scientists and researchers whose work depended on their understanding of how the Colossus or the UNIVAC operated, how to interpret the results it produced, and what to do if anything went wrong. Nonspecialists no more dreamed of understanding the workings of their computers than the average citizen expected to understand brain surgery or rocket science.

The economic success of home computers in penetrating the consumer market, however, was its appeal to computer users specifically as consumers. Purchasers of the PET and the TRS-80 did not need to know how computers worked or how to get under the hood to fix them. The great advantage of Apple computers in particular was that they could be operated by users who did not know anything about computers except how to run preinstalled programs. Although a hard core of tech-savvy computer geeks continued to tinker with their own hardware and software, Apple made it increasingly difficult for their purchasers to do so.

In the process, the original computer users—who had seen themselves as active participants in installing, maintaining, updating, and repairing their computers, along the lines of participants in a college seminar or postdocs in a laboratory—were redefined as students in a lecture course who absorbed information but were not asked to produce any of their own. Purchasers of home computers routinely became experts in using word processors to produce professional-looking documents or negotiating the obstacles in video games. But the goal of becoming more computer literate was merely a means to the end of running programs that would make it easier to do one's job or provide entertainment in one's leisure hours. As computer literacy was redefined from building to selecting a computer or from designing to running prepackaged software, computer users who were not running open-source software like Linux completed their transition from postdoc-style experts to undergrad-style amateurs.

This trajectory did not begin to be reversed until the gradual shift from Web 1.0 to Web 2.0 got under way with the emergence of the World Wide Web in the early 1990s and the rise of PDAs several years later. In 1989, Web inventor Tim Berners-Lee's proposal to marry hypertext to the Internet called for posting existing databases like telephone directories online by means of "a gateway program which will map an existing structure onto the hypertext model, and allow limited (perhaps read-only) access to it."[12] Berners-Lee went on to suggest the following year that the initial phase of preparing and networking a read-only online archive of texts, which he anticipated would take three months, be followed within another three months by enabling "the creation of new links and new material by readers. At this stage, authorship becomes universal."[13] The first files became available on the Web in 1991; by the beginning of 1993, five different browsers were available to pioneers wishing to surf it. Encouraged and enabled by the growing availability of broadband connections, users of personal computers could not only order an imposing range of goods and services online but also post and answer e-mail messages outside the network of their individual server, correct errors on reference sites like the Internet Movie Database (imdb.com), and share photos with friends to whom they specifically directed them or strangers around the world.

The rise of Wikipedia shows how far the pendulum had swung away from the user-as-consumer model that had made the personal computer a fixture in so many homes. Wikipedia's most distinctive advantages as a reference archive—its comprehensiveness, timeliness, and ceaseless improvement—depended on its ability to attract thousands of volunteers to create, expand, edit, and correct individual entries. Celebrating Wikipedia's relative freedom from restrictive rules on its use, expansion, or revision, Wales contrasted it with more restrictive websites whose protocols were designed to prevent malicious interference: "this philosophy of trying to make sure no one can hurt anyone else actually eliminates all the opportunities for trust."[14] By the time Wales wrote this, however, the Wikimedia Foundation had already felt obliged to institute rules and restrictions designed to promote minimal standards of civility and reduce the possibilities of deliberate defacing or the introduction of commercial material.

In pedagogical terms, Wikipedia might be described as combining elements of both Web 1.0 and Web 2.0, a lecture course and a college seminar. Wales describes Wikipedia as a quintessential instance of Web 2.0, a place where everyone's contributions are solicited on the assumption that their collective wisdom will be greater than the wisdom of even the most brilliant contributor. At the same time, each Wikipedia page contains elements of Web 1.0, a lecture hall designed primarily to communicate information effectively to the readers who consult it and limit the potential disruptiveness of unruly participants. Both attackers and defenders of Wikipedia have emphasized its similarities to free-wheeling seminars. But its affinities to lecture courses are equally important.

This third narrative, which locates Wikipedia toward the decentralized, libertarian end of a pendulum curve between defining computer users as consumers of products manufactured and distributed by more powerful and centralized corporations and empowering them as participants in an online community, ignores the lecture-hall quality of Wikipedia pages and protocols. Instead, it bases Wikipedia's claims to authority on its freedom from the centralized, hierarchical corporate and national interests that had directed the development of computer technology from Babbage's analytical engine to the rise of the Web. Instead of describing Wikipedia as the fulfillment of the wiki's promise, this apparently libertarian narrative turns out to be surprisingly conservative. It sees Wikipedia as a return to the earliest days of computing, before the ubiquitous black boxes on their desktops and laps turned the computing community into a collection of consumers isolated from each other and from the sources of computing power. The Wikipedia community, in this narrative, does not merely collaborate on the world's largest and most comprehensive wiki. In addition, its members return, with due nostalgia but on a once unimaginable scale, to the ethos of the World Wide Web Consortium, of Berners-Lee's workgroup at the European Organization for Nuclear Research (CERN), and of the collaborators on the Colossus project in World War II–era England.

It is no coincidence that the second and third of these groups cultivated a cooperative ethos in the service of military or strategic competition. For a foundational paradox of Wales's utopian vision of a human community based on mutual trust rather than the kind of antagonism that requires armed defense is that it does not inquire into the ultimate purpose of that community or the trust on which it depends. Every community that brings people together also separates them from members of other communities that have equally instrumental functions. Mutually shared interests are not universally shared interests; they are themselves instrumental. The corresponding paradox in liberal education has been unmasked by recent attacks on the myth that all parties to the academy share the same vision, the same sense of ends and means, the same ideas about the best ways to maintain or transform the current system.

Like Wikipedia in wanting to create a universal community devoted to the sharing of all human knowledge but forced by the misbehavior of some of its members to create and enforce unwelcome restrictions, the contemporary academy is caught in a series of contradictions. The oldest of these goes back to Socrates, whose ironic pose that he knows nothing leads to the Socratic questions that are still a staple of teaching in college classrooms as nowhere else. Graduate and undergraduate seminars meet around tables in spaces that proclaim every participant to be equally powerful, well-informed, and well-intentioned until it comes to approval, assessment, and advancement. It could hardly be otherwise, for teachers and students are not equal. Teachers not only have access to powers and perquisites students are seeking in and through their education; they are the gatekeepers who determine whether students will ever gain that access. Even more poignantly than the administrators of Wikipedia, academic administrators are sorely challenged to define students as acolytes, apprentices, consumers, or customers. Prospective employers, who increasingly prefer to outsource the costs of training employees for specialized positions, push for colleges to provide more vocational training, inevitably rendering liberal education more and more irrelevant. Contemporary debates about liberal education—from those in journals and monographs to those that play out every day in individual classrooms and offices—are largely driven by inabilities to resolve the resulting conflicts. Behind the unwelcome debate about how colleges should think of their students are further debates about what liberal education is for, and whether, like Babbage's analytical engine, its processes and products are reducible to a finite series of operations and points of information.

The Fourth Narrative: Wikipedia and Information Culture

The fourth narrative of Wikipedia's origins goes back the furthest of all: to the dawn of information depending on abstract categories, logical reasoning, and the ability to create, recognize, and decode symbols rather than gaining it through direct experience. Eric A. Havelock, Walter J. Ong, and James Gleick root this change in the shift from oral to literate cultures.[15] Gleick has traced the development of information culture from the invention and spread of alphabets through the production and archiving of manuscripts to the rise of encyclopedias. Both he and Reagle see prophecies of Wikipedia in earlier encyclopedias, to be sure. But they also see them in a more general "Enlightenment aspiration: a *universal* encyclopedic vision of increased information access and goodwill."[16]

As Andrew Brown has observed, even the oldest encyclopedic references share a vision of authoritative, organized knowledge with the successive revolutions in knowledge that produced, for example, the library, the codex book, and the scientific society. This vision, Brown contends, is derived from Aristotle's assumption that "one mind, or the mind, or mind, could grasp the whole of totality in a systematic and interrelated sweep." Bold as it is, this assumption leads inevitably to "the anxieties and contradictions that bedevil every encyclopaedia project." Some of the resulting paradoxes arise from the peculiar temporal status of encyclopedias, which Brown calls "monuments to transience," as both authoritative and always subject to revision. Some arise from the parasitism of encyclopedias on other sources: "if we are to trust in encyclopaedia, the information in it must be findable elsewhere: a work of reference needs to reflect (and copy) other texts." Some arise from the need to be both usefully specific in its parts and unified in its whole, a double requirement that allowed the compilers of the first edition of the *Encyclopaedia Britannica* (1769–71) to criticize its great competitor, Denis Diderot's *Encyclopédie,* as too atomistic in some respects and too thesis-driven in others. Some arise, surprisingly, from the alphabetical arrangement of encyclopedias, "an order that makes it easy to look things up" for any readers that can intuit the most relevant subject headings.[17] Although Ernst Cassirer has called alphabetical sequence "democratic," because it does not privilege any entry over any other, encyclopedias find all sorts of ways to reintroduce hierarchies, from the much greater length of some articles in all encyclopedias to Wikipedia's meta-entries that establish rules for all the other entries. The alphabetical paradoxes are in turn based on another paradox Brown derives from Alasdair MacIntyre's question: "What do you need to know before you can learn from 'great books'?" Every encyclopedia, however elementary, requires its users to know certain things (how to find alphabetically organized entries, how to use cross-references, how to read) before they can learn others. "How can they ever learn to use it?" Brown wonders of the autodidacts and slackers for whom Wikipedia and other encyclopedias substitute for universities: "What seems to be free for all is, on several levels, neither free nor for all." Finally, "the encyclopaedia is *the* work you are not supposed to quote verbatim. . . . The encyclopaedia says, 'I am the truth—but don't quote me on this'."[18]

It is no wonder that Brown calls Wikipedia "the '*Aufhebung*' of the encyclopaedic ideal—its consummation, its cancellation, and its raising to a higher level."[19] Apart from its freedom from the paradoxes of alphabetical sequence—although there are pages that list its entries alphabetically, most users never see these pages because they find the entries they seek by typing their headings directly into the site's search engine—Wikipedia inherits virtually all the paradoxes of authority common to all encyclopedias or indeed all archives. The authority of any encyclopedia or reference source depends on its being more comprehensive, better organized, and more readily searchable than its competitors. For years, sellers of the *World Book Encyclopedia* claimed that its division into volumes of unequal length, each collecting entries beginning with a single letter or a group of letters, made it easier to use than the *Encyclopedia Britannica,* whose equally sized volumes each began and ended arbitrarily. Wikipedia is far easier than either one to search for information on any given topic.

The real competition here, however, is not between *World Book* and *Britannica,* or between either of them and Wikipedia, but between any given encyclopedia and the sources it cites. Before the rise of the Web, the main argument for American households' purchase of print encyclopedias apart from their value as status symbols was their claim to be convenient and authoritative archives of a wide range of information in a single source. Encyclopedias could answer factual questions about science and history, provide general biographical information about a multitude of well-known figures, and give more detailed accounts of topics as varied as color, polar regions, and World War II. Even articles that did not entirely satisfy a given searcher's quest for information often included cross-references to other articles within the encyclopedia and bibliographic references that directed the searcher to still more detailed sources outside the document. Hence, encyclopedias, never the authorities of last resort, were marketed as the best places for librarians, householders, and students to begin their research, places that often but not always made it unnecessary to search further. These are exactly the same claims both attackers and defenders have made about Wikipedia: that, even though it is not the ultimate authority about anything, it is the first authority many researchers consult precisely because it is so easily accessible.

All encyclopedias, print or online, are by their nature secondary or tertiary sources whose authority is not based on any original research their contributors have done but on their synoptic range and organization. Although the long shelf of hefty, uniform volumes that constitute most encyclopedias may seem to project a massive, undifferentiated authority, the authority encyclopedias have is actually of a very particular kind, for an encyclopedia is not a collection but a digest of knowledge. What gives the contributors to encyclopedias their authority is not their eminence as researchers, though that incidental strength has sometimes been advertised as primary, but their knowledge of a given field and of where to find reliable information about it and their ability to write in a selective and accessible way that presents the facts most likely to be useful to the readers most likely to consult their articles. No encyclopedia is equally authoritative throughout; some of its articles are more authoritative than others. Despite the editorial hand that seeks to impose a degree of uniform quality control in every print encyclopedia, the editors who wield this control cannot, by another paradox of authority, possibly be such experts in every field an encyclopedia covers that they can authoritatively overrule their contributors on matters of fact or emphasis. The most they can hope is to impose a more homogenized voice free of stylistic blemishes and obvious howlers. The attenuation of central editorial control, widely noted in both attacks and defenses of Wikipedia, merely emphasizes paradoxes already unavoidably present in print encyclopedias—paradoxes to which I now turn closer attention.

CHAPTER TWO

Paradoxes of Authority

WIKIPEDIA'S ORIGIN STORIES are not the only feature it shares with comic book superheroes. Another is its power to unmask its sometime antagonists in the encyclopedia business and the higher-education business by throwing into sharper relief the paradoxes behind their claims to authority. These paradoxes have been implicit in academic authority for many years, but Wikipedia's accounts of its own origins invite us to consider them anew. Whatever the authority of Wikipedia articles on the Arab-Israeli conflict or Barack Obama citizenship conspiracy theories or of Wikipedia as a general project, examining more closely the claims to authority that both Wikipedia and the academy make can lead to even more surprising and illuminating insights into the problematic nature of all authority.

Authority in Wikipedia

Despite its status as the quintessential example of Web 2.0, Wikipedia makes no bones about its claims to authority. The article "Wikipedia: Why Wikipedia Is So Great" trumpets its comprehensiveness (four and a half million articles and counting, compared with the *Encyclopaedia Britannica*'s 65,000), its freedom from the restrictions on length common to all print references, its neutrality, which makes it "an excellent place to gain a quick understanding of controversial topics," its openness to editing by anyone ("even if you're too young to legally tell us your name"), the timeliness of articles that are updated "thousands of times an hour," the swiftness with which revisions can be made ("Errors to Wikipedia are usually corrected within seconds, rather than within months as it would be for a paper encyclopedia"), and its resistance to vandalism ("Wikipedia, *by its very nature,* resists destructive edits").

The broad implication of this article is that Wikipedia's greatness as a reference source rests ultimately on two advantages it has over competing authorities: the freedom it accords its users ("Wikipedia is free") and especially its editors ("Wikipedia has got 'consensus seeking' down to a fine art") and its ability to attract "highly intelligent, articulate people (with the exception of repeat vandals) with some time on their hands. . . . The Wiki-community of Wikipedians is a group of special people who have special characteristics." In other words, Wikipedia is a radical democracy that just happens to comprise an elite group of participants, "highly intelligent people . . . from all over the world."[1]

This article in particular, and Wikipedia in general, is well aware of these contradictions. Although they are never resolved, they are contained and managed by several strategies. The article is clearly labeled an "essay" that "contains the advice or opinions of one or more Wikipedia contributors. Essays may represent widespread norms or minority viewpoints." Readers are advised to "consider these views with discretion. Essays are not Wikipedia policies." Its expansive claims for the greatness of Wikipedia are balanced by the analysis of the inevitable counterarticle to which "Why Wikipedia Is So Great" is linked: "Wikipedia: Why Wikipedia Is Not So Great." The first article moderates its claim that "Wikipedia has almost no bureaucracy; one might say it has none at all" with the qualification, "but it isn't total anarchy. There are social pressures and community norms, but perhaps that by itself doesn't constitute bureaucracy, because anybody *can* go in and make any changes they feel like making." According to this account, the ability of any user to edit Wikipedia, and to edit anyone else's edits, allows Wikipedia to avoid the problems of both bureaucracy and anarchy. "Wikipedia: Why Wikipedia Is Not So Great," which "began in 2001 as a humorous numbered list of issues, edited into serious tone in January 2004" (and to which "WP: SUCKS" redirects), begs to differ: "Despite claims to the opposite, Wikipedia is a bureaucracy, full of rules described as 'policies' and 'guidelines' with a hierarchy aimed at enforcing these (sometimes contradictorily)."[2]

Anti-authoritarian Authority

As Wikipedia cofounder Jimmy Wales acknowledged when he was interviewed in the film *Truth in Numbers?*, Wikipedia "really is not a democracy. . . . There's a whole internal structure to it."[3] Within this organizational structure, some members have more power than others. At the bottom are the innumerable mostly anonymous or pseudonymous users who collectively make thousands of changes to Wikipedia pages per minute. Users who have been registered for a Wikipedia account for at least four days and have made at least ten edits during that time become autoconfirmed editors with the power to move Wikipedia articles, edit semi-protected articles, and vote in certain elections.

More elite than the autoconfirmed editors are 1,409 administrators ("sysops"—system operators—or "admins") who have been "approved by the community,"[4] whose status dates from a memo from Wales dated 11 February 2003, which announces that "I think perhaps I'll go through semi-willy-nilly and make a bunch of people who have been around for awhile sysops" and emphasizes that "becoming a sysop is *not a big deal*": "I want to dispel the aura of 'authority' around the position."[5] Among the powers administrators have is the ability to edit protected pages, view and restore deleted pages, hide and delete page revisions, restrict pages from editing, edit Wikipedia's main page, and block specific users from editing. Any Wikipedia user may apply to be an administrator. Applications are posted for seven days on the WP: RfA page, where any editor can read, comment, and vote on them. Acceptance is based on consensus rather than majority approval; "most requests above ~80% approval pass and most below ~70% fail"[6]—a statistic that hints at just how few applications for administrator status are approved.

Despite Wikipedia's insistence that it is not a bureaucracy, administrators are distinct from a much smaller corps of bureaucrats, who currently number thirty-five, not counting twenty former bureaucrats who resigned or were removed. Bureaucrats are given the power (some might say the authority) to "add the administrator, bureaucrat, bot [an automated program design to handle routine tasks or edits], account creator, or reviewer user group to an account," to "remove the administrator, bot, account creator, IP block exemption, or reviewer user group from an account," to rename accounts, and to promote users to the status of administrators or bureaucrats." "Bureaucrats are not super-admins," but they apply for bureaucrat status in a similar way, and it is granted according to a similar process, although "the expectations for potential bureaucrats are higher and community consensus must be clearer."[7]

Over the bureaucrats comes the Arbitration Committee, whose formation Wales announced in December 2003. Wales wished to delegate the substantive disputes among editors, determinations of which editors have the "CheckUser" power to check IP addresses of suspected vandals and the oversight power of "enhanced deletion" over libelous material or material that infringes copyright, and decisions about banning users that he had heretofore made on his own. In 2004, Wales appointed twelve committee members from among editors who had volunteered to help with mediation and arbitration. Since then the committee, "which is analogous to Wikipedia's supreme court,"[8] has been expanded to include eighteen members who are currently appointed for staggered two-year terms after annual advisory elections.

Standing apart from the Arbitration Committee are the stewards, whose number is not limited (they currently number thirty-eight) and who have been elected "once to twice a year . . . by the global Wikimedia community," since April 2004. Stewards, who are "tasked with technical implementation of community consensus," might be considered "global sysops" who have "complete access to the Wiki interface on all Wikimedia wikis, including the ability to change any and all user rights and groups."[9]

The "ultimate corporate authority"[10] for the Wikimedia Foundation is not the bureaucrats, arbitrators, or stewards, but the Wikimedia Board of Trustees, founded in 2003 with three members and expanded over the years to ten. This board manages the foundation and its budget. The foundation's stated purpose is "to empower and engage people around the world to collect and develop educational content under a free license or in the public domain, and to disseminate it effectively and globally."[11] Trustees are elected to two-year terms, with the exception of the community founder trustee, a position reserved for Jimmy Wales, who is reappointed from year to year (if he were not reappointed, the position would remain vacant). Joseph Michael Reagle Jr. has aptly noted that Wales has defined his own position as more like that of a constitutional monarch than that of a benevolent dictator. Indeed it is "a hybrid of leadership types including autocratic (decision made by the leader alone), consultative (the problem is shared with and information collected from the group, before the leader decides alone), and delegated leadership (the problem is shared, ideas are accepted, and the leader accepts the solution offered by the group)."[12]

This summary of Wikipedia's administrative structure supports several conclusions drawn by Mathieu O'Neil. Wikipedia does indeed have an administrative structure that represents, in O'Neil's terms, "*a combination of democracy and bureaucracy*."[13] This structure has gradually evolved from a benevolent despotism through Wales's desire to delegate powers he had formerly wielded himself. Wikipedia's administrative structure, which looks so ad hoc and self-selecting at the bottom tiers, looks much more corporate at the top. Wikipedia depends on a "radical redefinition of expertise, which is no longer embodied in a person but in a *process*: the aggregation of many points of view."[14] Since "leadership on Wikipedia is intimately associated with the person and beliefs of Jimmy Wales," as O'Neil points out, "Wikipedia exhibits the opposite pull of the two principle [*sic*] forms of legitimate power in online tribes, charismatic and sovereign authority"[15] that derive respectively from the will of strong leaders and the will of the people who follow them. Finally, the structure is marked, especially at the lower levels, by Wales's allergic reaction to the terms "bureaucracy" and "authority," even though the structure includes a group of regulators specifically named bureaucrats, even though authority is exactly what it is designed to maintain and regulate. As Wikipedia cofounder Larry Sanger says in *The Truth According to Wikipedia*: "The notion that there would be anyone in authority . . . was just completely anathema to them."[16] The systematic disavowal of its own authority, in fact, is central to the way Wikipedia is conceived and organized.

Levels of Equality

Users, editors, administrators, and Jimmy Wales may not agree whether Wikipedia's administrative structure amounts to a bureaucracy whose different offices exercise distinct kinds of authority, but they all agree that not all Wikipedia articles have been created equal. At the top of the pyramid are 3,943 "featured articles" (about one out of every 1,120 articles in the English-language Wikipedia), each marked by a bronze star, that have been approved by Wikipedia's editors as "distinguished by professional standards of writing, presentation, and sourcing," because, in addition to following Wikipedia's guidelines for structure, length, and illustrations, they are "well-written . . . comprehensive . . . well-researched . . . neutral . . . [and] stable."[17]

Just below featured articles are 18,127 "good articles" (about one in 237 Wikipedia articles in English) the editors have approved and marked by a circled green plus sign. Like the featured article to which it aspires to become (and to lose its label as a good article through the promotion), a good article is "well-written . . . verifiable with no original research . . . broad in its coverage . . . neutral . . . stable . . . [and] illustrated, if possible, by images."[18] The main difference between these criteria and those for featured articles is the difference between the comprehensiveness required of every feature article, "which neglects no major facts or details and places the subject in context," and the breadth expected of a good article, which "addresses the main aspects of the topic." Many short articles, or relatively brief overviews of broad topics, that would not meet the criteria for featured articles have still been designated good articles.

It is easy to tell the difference between good articles and what might be labeled run-of-the-ordinary Wikipedia articles by considering several "quickfails" that can eliminate a nominated article from promotion to a good article. It may lack reliable sources. It may treat its topic in a way that violates Wikipedia's demand for a neutral point of view. It may contain cleanup banners that ask users to supply further references, citations, or clarifications. It may be the subject of ongoing or recent edit wars conducted by opposed partisans who fight back and forth to determine what a given article should say. Its subject may concern current events whose determinate end has not yet been reached. It may be suspected of close paraphrases or copyright violations of its sources.

Wikipedia often makes distinctions even among articles that have not been rated featured or good. Among its thousands of articles on American primary and secondary schools, for example, articles are rated on a quality scale as A-Class (if they meet the "good articles" criteria), B-Class, C-Class, Start-Class, or Stub-Class. On the separate scale of importance, articles may be rated Top, High, Mid, Low, or ??? (Unknown). Articles that consist exclusively or primarily of lists of schools have their own metrics, including Featured List and Stand-alone List.[19] The great majority of articles, however, carry no markers of assessment.

Clearly, some Wikipedia articles, like some Wikipedia users and editors, are more equal than others. But Wikipedia is much more reluctant to acknowledge different kinds or levels of authority among its editors than among its entries, because any admission that some users and editors wield more authority than others would undermine, or at least complicate, Wikipedia's proud claim to be the people's encyclopedia, or, as its home page proclaims, "the free encyclopedia anyone can edit."[20] Wikipedia's passionate devotion to the ideals of freedom and democracy, which may seem on an unavoidable collision course with its claims to authority, reveal on closer examination the paradoxical nature of those claims.

The most obvious of these paradoxes is buried in the assumption *vox populi, vox dei*—the voice of the people is the voice of God. The people as a mass may be the most powerful force in any democracy, but they are not necessarily the most intelligent. Indeed the whole presumption of representative democracies is that the people need more than simply access to the workings of government and an active voice in its operation; they need leadership from representatives who advocate for their views but also proceed with greater deliberation and wisdom than a mob would. When the people's voice is chosen as the sole authority, it can produce the *Family Feud* effect, endorsing the most popular answers over the correct ones. In a world in which 25 percent of all Americans believe that U.S. armed forces found weapons of mass destruction in Iraq, the reliability of crowdsourced authority is fraught with dangers.

Wikipedia's current administrative structure arose directly out of the soul-searching that followed the 2005 creation of a false and defamatory entry on John Seigenthaler by Wikipedia user Brian Chase. Chase's motivation was telling: he posted the information as a prank on a friend who knew Seigenthaler as a local celebrity who had formerly edited the *Tennessean* and created its First Amendment Center. Chase told *USA Today,* where Seigenthaler had been founding editorial director, "I didn't think twice about just leaving it there because I didn't think anyone would ever take it seriously for more than a few seconds." More to the point, Chase did not take Wikipedia seriously himself, considering it "some sort of 'gag' encyclopedia."[21] Wales's hope was that, just as Chase had felt free to deface Wikipedia because he did not take it seriously, users who did take it seriously, in part because of the administrative changes Wikipedia made in response to the Seigenthaler incident, would not act so irresponsibly. For better or worse, Wikipedia's reputation for authority depends directly on the activities and attitudes of its users. The more users treat Wikipedia as either a gag or a serious reference, the more like that it becomes.

Some sources might seek to bolster their authority by requiring their contributors to sign their articles, allowing readers to review their credentials and judge their claims to authority. Indeed this was exactly the issue over which Wikipedia's other founder, Larry Sanger, says he split with Wales, who refused to distinguish between expert and nonexpert contributors. After leaving Wikipedia, Sanger launched the rival online encyclopedia Citizendium, whose scope remains far more limited than that of Wikipedia but whose articles are granted "approved" status only if they stand up to the examination of designated experts. Article 4 of Citizendium's charter makes this departure from Wikipedia explicit: "The Citizendium community shall recognize the special role that experts play in defining content standards in their relevant fields and in guiding content development towards reliability and quality."[22] The credentials of these experts can vary widely, as users can see from comparing the credentials of the editors who approved the pages on Richard Hofstadter, Joe Louis, and Doom (video game). But they are all identified by name.

Ironically, Citizendium's insistence on recognizing the special role of experts and identifying its experts by name revealed further problems of authority. As the RationalWiki article on Citizendium contends:

> There's expertise and then there's certification as an expert, which is a social construct made of pieces of paper and (hopefully) accredited standards. Sometimes the two don't quite overlap. In the quest for expertise—"This

> article is good and I can explain why"—Citizendium went for credentialism—"This article is good because I have the authority to say so." . . . But the most damaging part of the Citizendium approach was that the required credentials are inconsistent. Someone wishing to be a general editor in an academic field must prove they have a PhD, or are a tenure-track professor. . . . But if they wanted the authority to take over articles in alternative medicine, they only had to prove that they were licensed to practice their branch of alternative medicine.[23]

Once advocates of homeopathic medicine were credentialed and protected as experts in their field, the result was a flood of articles endorsing homeopathic medicine duly approved by these experts, who successfully repelled all attempts by mainstream scientists and physicians to question their authority.

Wikipedia, by contrast, values the anonymity of its contributors so highly that it allows editors, administrators, bureaucrats, members of the arbitration committee, and stewards to post online only usernames that keep their real names secret. Since it does not oblige contributors to identify themselves, it bolsters its authority by appealing instead to sources outside itself. Contributors are enjoined to add citations and bibliographies to the pages they create and edit; pages with a critical mass of such citations are more likely to be awarded the coveted status of good article or featured article; and warnings are posted on pages the editors deem deficient in such references.

Three Paradoxical Principles

The paradoxes implicit in Wikipedia's attempt to devise nonauthoritarian structures of governance and levels of reliability are mirrored in its core principles governing its content. The page "Wikipedia: Core Content Policies" sets forth three policies governing all content posted to Wikipedia: material must be presented from a "*neutral point of view,* representing significant views fairly, proportionately and without bias"; it must be verifiable—writers and editors must provide reliable print sources for everything they post on Wikipedia so that users can verify the accuracy of its information—and none of it may be based on original research, but only on material available elsewhere. Each of these principles turns out to be equally paradoxical—not because of Wikipedia's susceptibility to logical errors, but because of the paradoxical nature of authority itself.

PARADOXES OF VERIFIABILITY

The most obvious of these paradoxes, stemming from the need for verifiability, is familiar to editors and reviewers of all reference volumes, from dictionaries to almanacs: aspiring authorities regularly seek to establish their preeminence among competing authorities by plundering those very authorities. As I suggested in chapter 1, this is the fundamental paradox enshrined in Newton's phrase "on the shoulders of giants": scientists, researchers, and creators best acknowledge the importance of their predecessors by making new contributions that successfully compete with the work of those predecessors, excelling and perhaps even superseding them. So the search for authority is ultimately a search to contest authority.

This paradox is complicated by an awareness of different modes of authority. Footnotes to contemporaneous historical sources would do little to bolster the authority of the four Gospels' accounts of the life and work of Jesus, which claim a very different kind of authority from a kind that footnotes could support. In the same way, the endnotes Wikipedia craves in order to root its authority more firmly might well end up rendering the very idea of authority more problematic. At the least, they raise more questions than they answer about hierarchies of authority. At the heart of Wikipedia's battles against its critics and its own presumably more imperfect incarnations of 2001, 2009, or yesterday is a series of questions about the nature of authority. What makes an expert authority on a given subject an expert, and how are experts certified and recognized? What can researchers do when the experts they consult disagree? Are there hierarchies of more and less expert experts, and who has the power to constitute such authorities? What forces make hierarchy subject to change, and how can researchers judge the legitimacy of these changes? If experts are certified only by other experts, is it possible or desirable to break out of the circle of experts whose claims to authority are radically and necessarily interdependent?

An invaluable experience Wikipedia makes available to anyone who is seriously interested in critically examining the nature of authority is the opportunity to consider questions like those above, questions that are patently ignored by fledgling researchers like high school students, frequently disavowed even by acknowledged experts in given fields, and systematically effaced by traditional encyclopedias and often by the broader institutions of print culture. Although its paradoxical status as the people's encyclopedia cannot help raising these questions, Wikipedia attempts to finesse them by framing its search for authority in strategically limited terms. Its goal is "verifiability," not truthfulness: "Material challenged or likely to be challenged, and all quotations, must be attributed to a reliable, published source."[24]

This position reduces the different sorts of authority with which Wikipedia might have to deal to only two: print publication in a reliable source, which is authoritative, and any other source, which is not. Private citizens and public figures who wish to add information or correct errors on pages devoted to themselves cannot do so unless they can cite print sources for the new or corrected information, as I discovered when a colleague of mine, urged by his students to expand and correct his own Wikipedia page, found his edits reverted as inadequately sourced. He was not considered a reliable source about his own life and credentials according to Wikipedia's guidelines, which rule that articles should be based on "reliable, third-party, published sources with a reputation for fact-checking and accuracy. . . . The best sources have a professional structure in place for checking or analyzing facts, legal issues, evidence, and arguments. The greater the degree of scrutiny given to these issues, the more reliable the source. . . . Where available, academic and peer-reviewed publications are usually the most reliable sources."[25]

This formulation makes it clear that Wikipedia's refusal to anoint some contributors or editors as more authoritative than others amounts to a decision to leave that decision to others—specifically, to the gatekeepers of published, ideally academically reviewed, sources like this book. By deliberately positioning itself as a compendium of the knowledge to be found in reliable printed sources, Wikipedia makes its claims to authority much easier to defend. By leaving all further decisions about authority to the sources to which it defers, however, it drastically limits all such claims to whether the information it includes is verifiable, not whether it is true. Unlike Newton, Wikipedia does not claim to excel its sources. Indeed, despite its boasts about its breadth and comprehensiveness as a global enterprise, it is a matter of policy that no specific article should seek to improve on its sources.

Verifiability, which was established as the second of Wikipedia's three core content policies in August 2003, arose as a way to hold contributors to the standard established by the first policy, the need to present all information from a neutral point of view. The rationale behind the NPOV policy, which dates from February 2002, is that if contributors with different, often conflicting, beliefs are to collaborate productively, they must operate from a base of shared beliefs, and "while it is often hard for people to agree as to what is the truth, it is much easier for people to agree as to what they and others *believe* to be the truth. Therefore, Wikipedia does not use 'truth' as a criterion for inclusion. Instead, it aims to account for different, notable views of the truth."[26]

As verifiability was designed as a corollary to NPOV, Wikipedia's third core principle, "no original research (NOR)," was codified as a response to tendentious contributors who

> would often marshal sources to argue that a minority view was superior to a majority view—or would even add sources in order to promote the editor's own view. Therefore, the NOR policy was established [later] in 2003 to address problematic uses of sources. The original motivation for NOR was to prevent editors from introducing fringe views in science, especially physics—or from excluding verifiable views that, in the judgement of editors, were incorrect. It soon became clear that the policy should apply to any editor trying to introduce his or her own views into an article (and thus a way to distinguish Wikipedia from Everything2).[27]

This revealing final parenthetical phrase shows Wikipedia's determination to establish itself as an encyclopedia as distinct from the proudly iconoclastic website Everything2, which includes not only "informative articles on just about any subject you can imagine—and many that you never thought of"—but also "humor, poetry, fiction, opinion, criticism, personal experiences, and other things that are hard to categorize."[28]

PARADOXES OF THE NEUTRAL POINT OF VIEW

The quest for a neutral point of view in every Wikipedia article is laudable but quixotic, even disingenuous. As Larry Sanger has pointed out, the term NPOV "implies that to write neutrally, or without bias, is actually to express a *point of view,* and, as the definite article is used, a *single* point of view at that."[29] Since every contributor is likely to assume that his or her point of view is unbiased, a truly neutral point of view, as "Wikipedia: Core Content Policies" points out, can only be operational, not substantive. Even if the neutral point of view is one that represents all important viewpoints, someone must decide which viewpoints are important enough to include in a given article, and which of the important viewpoints deserve the most prominent representation.

The case of Holocaust deniers is instructive. The original 2001 entry on the Holocaust devoted a brief section to "Revision and Criticism," a section that originally read in toto, "Some Neo-Nazi groups, and others, have sought to deny that the Holocaust ever occurred, or to sanitize it. These revisionist views are rejected by all serious historians of the period. See Holocaust revisionism for details."[30] This section was repeatedly revised and expanded until it was moved the following September to its own entry. The only place the phrase "Holocaust denial" appears in the lengthy current Wikipedia entry on the Holocaust is as one of a list of ninety-four hotlinks appended at the very end of the article. Instead of giving Holocaust deniers a place in the Holocaust article, the editors chose to give them their own article, treating the phenomenon of Holocaust deniers, if not their beliefs, as a historical fact quite distinct from the Holocaust. But unlike historians of the Holocaust, its deniers must share their article with their critics, who make a prominent appearance as early as the third paragraph: "Most Holocaust denial claims imply, or openly state, that the Holocaust is a hoax arising out of a deliberate Jewish conspiracy to advance the interest of Jews at the expense of other peoples. For this reason, Holocaust denial is generally considered to be an antisemitic conspiracy theory."[31] Much of the rest of the article on Holocaust denial, which is almost as long as the article on the Holocaust, returns to this censorious tone, which presents the denial of the Holocaust as a fringe opinion in either venue. The tone of both articles may seem reasonable and appropriate. But just as Conservapedia advances unanswered criticisms of Democratic policies in its article on the Democratic Party but omits or marginalizes criticisms of Republican policies in its article on the Republican Party, Wikipedia's articles on the Holocaust and Holocaust deniers represent a distinct point of view that is neutral only in the specific sense it aims for: it accurately represents consensual thinking about the Holocaust.

PARADOXES OF NO ORIGINAL RESEARCH

The third and most interesting of Wikipedia's core principles is its proscription against original research. According to Wikipedia's NOR policy, "all material added to articles must be *attributable* to a reliable published source, even if not actually *attributed*." Even the most commonsensical assertions, those for which it would be pointless to add footnotes or online citations, must be accessible in some print source. But "despite the need to attribute content to reliable sources, *you must not plagiarize* them or violate their copyrights. Articles should be written in your own words while substantially retaining the meaning of the source material."[32] Wikipedia contributors must therefore walk a tightrope between plagiarism and the addition of original ideas or interpretations, both of which are expressly forbidden.

Wikipedia distinguishes between primary sources like eyewitness historical accounts, records of scientific experiments, or original works of literature; secondary sources like political or scientific or literary histories that interpret primary sources; and tertiary sources like encyclopedias and other reference works that summarize the findings of secondary sources. It warns fledgling researchers tempted to cite it as a reliable source that it differs from "other encyclopedias, such as *Encyclopaedia Britannica,* [which] have notable authors working for them and may be cited as a secondary source in most cases."[33] Wikipedia never claims the status of a secondary source, but defines itself as a tertiary source that may draw to a limited extent on primary sources or other tertiary sources but that depends on material mainly and properly drawn from secondary sources, so that "an article about a novel may cite passages to describe the plot, but any interpretation needs a secondary source. *Do not* analyze, synthesize, interpret, or evaluate material found in a primary source yourself; instead, refer to reliable secondary sources that do so."[34]

In a post dated 3 December 2004, Jimmy Wales explained that the NOR policy was designed "as a practical means to deal with physics cranks" by waiving the question of whether "someone's novel theory of physics is valid," deferring to the judgment of "reputable publishers" instead: "it's quite convenient to avoid judging the credibility of things by simply sticking to things that have been judged credible by people much better equipped to decide."[35] There are several problems with this convenient position. The first is implicit in Wales's own illustrative example, gun-control advocate Michael Bellesisles's *Arming America: The Origins of a National Gun Culture*, which uses an examination of county probate records to argue that only 14 percent of American households owned a gun before the Civil War. This information so surprised pro-gun scholars that they decided Bellesisles's research must be faulty and proved to their satisfaction that it was. An investigation into Bellesisles's own research led to the withdrawal of the Bancroft Prize Bellesisles had been awarded in 2001 and his resignation from Emory University. Wales notes that Wikipedia would "quite properly have rejected" Bellesisles's "novel historical thesis . . . because we are ill-equipped to judge the validity of such things." What he fails to note is that deferring to the authority of Bellesisles's publisher, Alfred A. Knopf, or the first round of reviewers who praised his work in scholarly journals would not have kept Wikipedia out of trouble if it had been around to compile an article on Bellesisles in 2000. The lesson of the Bellesisles affair is merely that outsourcing questions of authority to secondary sources gives Wikipedia deniability without necessarily making its entries any more reliable.

Another problem with the NOR policy emerged in the online response to Wales's post from concerned members of the Wikipedia community. How could Wikipedia editors distinguish between the synthesis of material from secondary sources, which the prohibition against plagiarism seemed to require, and the sort of original research that was forbidden? A debate over the policy among Wikipedia editors is revealing. One respondent, Zoney, argues that "valid Wikipedia articles can indeed be considered 'original research' by others outside Wikipedia, as articles should arise from the gathering of information from various sources (a process called research?) with the result of an article unique to Wikipedia."[36] But Mark Richards replies, "An article that makes no new low-level claims, but nonethless [*sic*] synthesizes work in a non-standard way, is effectively original research that I think we ought not to publish."[37]

After Wales writes in support of this latter position, Shane King points out still another complication: "We haven't solved the problem. We've shifted the burden of evaluating the credibility of the theory to evaluating the credibility of the sources. I see no reason to believe we're any better at evaluating the credibility of sources than of theories." He adds: "I largely see the NPOV policy and the 'no original research' policy as being in conflict. We have to report neutrally on all views, yet we exclude views that experts don't deem credible."[38] Wales, ignoring the lesson of Michael Bellesisles, responds that "it is a lot easier to evaluate the credibility of sources than the credibility of theories. If you offer me your personal theory of 'Liquidity, Efficiency, and Bank Bailouts' then it's going to be quite hard for me to judge whether you are an economics crank or someone with an interesting theory. But if you point me to an essay of that title in *American Economic Review,* I can feel comfortable that it is at least credible." He dismisses King's charge that the NOR policy excludes points of view that have not been endorsed by experts: "Sometimes we exclude views, but more commonly we move them to where they belong—in an article about theories that are not widely accepted."[39] As the example of the Holocaust and Holocaust denial articles demonstrates, however, sequestering minority views in their own articles merely muffles conflicts rather than guaranteeing neutrality in their presentation.

King argues in response that the NOR policy is not only unhelpful but unnecessary, since its most important effects are achieved by the NPOV policy. He asks, "How are we to write articles about pseudoscientific topics, about which majority scientific opinion is that the pseudoscientific opinion is not credible and doesn't even really deserve serious mention?" His summary could be applied equally well to Wikipedia's articles on the Holocaust and Holocaust deniers:

> The task before us is not to describe disputes fairly, on some bogus view of fairness that would have us describe pseudoscience as if were on a par with science; rather, the task is to represent the majority (scientific) view as the majority view and the minority (sometimes pseudoscientific) view as the minority view, and, moreover, to explain how scientists have received

pseudoscientific theories. This is all in the purview of the task of describing a dispute fairly.[40]

Given the fact that, as Jay JG points out, that "proponents of the various fringe theories [who] insist that their trivia is crucial to understanding the issue" are "much more determined about their obsession, er, area of focus, than the average editor,"[41] it is hard to see how anyone can take a neutral view of what counts as a reliable source without doing the original research that would justify that claim: "Partisans on different sides of an issue will insist that the sources brought by the other side are 'biased' or not reputable. Is CNN reputable? Fox? Al Jazeera? The *Jerusalem Post*?"[42]

Mark Richards notes that, since proponents of fringe theories often claim that "the 'credible' scientific sources are involved in some kind of conspiricy [*sic*], or are systematically unable to appreciate the field for some reason," it is impossible to avoid "picking and choosing which sources are 'credible' based on what we believe to be right."[43] Ray Saintonge caps this argument by observing, "If we 'are picking and choosing which sources are "credible" based on what we believe to be right' then the debate has just been shifted to one of determining what we mean by 'we.' The debate has not come to a resolution; it has merely shifted its focus. The 'we' that participates in a debate is not the same 'we' that takes one side of a particular argument."[44] The prohibition against original research ends by replacing orthodox beliefs with what literary theorist Stanley Fish, writing from a comfortable distance from Wikipedia, would call an "institutional community"[45] that has already decided what it consensually believes.

Still another paradox involves the question of whether passionate interest in a particular problem amounts to expertise about it. Wales defends an earlier formulation of Wikipedia's NPOV policy's passage that asserted, "If we are to represent [any given] dispute fairly, we should present competing views in proportion to their representation among experts on the subject, or among the concerned parties."[46] But Richards observes that "concerned parties" are distinct from "experts on the subject" and often adopt a tendentious tone in writing and editing that is antithetical to expertise because it emphasizes their bias: "When I read an article and it sounds to me like whoever wrote it is trying to push a point of view, it irritates me, even if I agree with the point of view. Think, 'does this sound like it's written by someone with strong feelings on the subject?,' and if so, why, and how can we change that?"[47] In this view, being a concerned party to a dispute would tend to disqualify rather than qualify an observer as a contributor to an article touching on that dispute.

Joseph Michael Reagle Jr. compares the notion of *credibility* under debate in this discussion to the *confidence* with which many scientists endorse theories of global warming. "There is evidence that determines our confidence in both facts and theories," he observes, and adds: "With respect to theory, confidence is determined by the nature of its assumptions, testability, the quality of the underlying data/observations, and the theory's explanatory power. Theories are wrestled with by the scientific community, tested, repeated, confirmed, and settled upon by scientific consensus and ultimately judged by historical hindsight."[48] As Reagle acknowledges elsewhere, however, "scientific does not equal academic: academic authority is based on a hierarchical application of judgment to those who allegedly know less; while closely associated with the academic, scientific assessments should be discernible to those who know the same or even less."[49]

One might argue that since no one has the patience, energy, or perversity to subject every possible theory—global warming, evolution, gravity—to the kind of testing that would produce verifiable results, everyone ends up punting on questions of authority, delegating judgments to someone who is stipulated as a higher authority. In this view, there is nothing unusual about the problems Wikipedia's core principles raise, for they are hardly unique to Wikipedia. Reagle's distinction between scientific and academic authority, however, suggests several illuminating implications the way that Wikipedia has raised and managed these problems has for the liberal arts—that is, for those areas of liberal education and research that are not governed by scientific principles.

OTHER MODELS OF NO ORIGINAL RESEARCH

Consider the models of authority Wikipedia's prohibition of original research most closely resembles. The model most often invoked by Wikipedia is the *Encyclopaedia Britannica* and other tertiary sources. But Wikipedia departs from *Britannica,* and proudly so, in many ways—for example, by including on the opening page of its English-language edition a summary of recent news stories that supplement its aspirations as a repository of all human knowledge by emphasizing its timeliness, even if the changing current events summarized each day on this opening page are of fleeting interest. So it makes sense to consider other models of authority that resemble Wikipedia more closely in one way or the other.

One such model is *Reader's Digest,* which, like Wikipedia, aims to make available information originally published in reputable sources rather than original material. Although *Reader's Digest* now publishes unabridged digital editions for the iPad, Kindle, and Nook, and much of its material is now available at www.rd.com, its roots in hard-copy publication distinguish it from Wikipedia. So does its approach to that material—licensed abridgment rather than unlicensed summary and fair-use quotation—and its assumptions about what counts as reliable sources. Even so, *Reader's Digest* offers its readership the same fundamental guarantee as Wikipedia: there's nothing new here, nothing you couldn't readily find elsewhere if you were willing to invest the time and effort. Indeed, *Reader's Digest,* which includes humorous anecdotes and first-person narratives previously unpublished, comes closer to including the products of original research, or at least undocumented personal experience, than Wikipedia does. Although the accuracy of its summaries of particular articles can readily be established by comparing them to their sources, the only authority offered for the most important claim of their shorter anecdotes—that they are true—is the word of the unfamiliar authors who have submitted them.

Wikipedians who bristle at the suggestion that Wikipedia is a digest would doubtless be even more offended to have it compared to the term papers students are asked to write for high school courses. Yet the similarities here are even deeper. In neither case is professional subject expertise or credentials in writing and editorial practice required. Writers in both cases are expected to negotiate between the contrary perils of original research and plagiarism in order to produce essays that draw on published material without claiming the authority of that material. The writing of term papers, of course, is an apprentice's exercise designed as an early stage in the acquisition of expertise whose product, the finished paper, is no more than a byproduct of the learning process. This obvious difference between term papers and Wikipedia, however, is blurred by the encouragement Wikipedia offers teachers to ask students to edit individual entries as class assignments and by the fact that every Wikipedia entry, because it is subject to endless revision by an expanding community, is a work-in-progress that is as much process as product itself.

Wikipedia may be more like Fox News, which presents itself in all its incarnations as "Fair & Balanced."[50] The similarity here turns not on the question whether either Fox or Wikipedia is indeed fair and balanced but on their determination to present themselves in those terms. Wikipedia's NPOV and NOR policies are textbook instances of Fox's slogan: "We report, you decide." Ongoing debates about the ability of either institution to live up to these slogans reinforce rather than undermine the similarities between the two.

A more suggestive analogy than any of these is the practice of textual adaptation that turns comic strips into blockbusters, theatrical plays into musicals or operas, novels into plot summaries, and summer movies into video games. For most nonprofessional observers, the cardinal virtue of adaptations of novels by Jane Austen or J. K. Rowling, or of comic strips from *Batman* to *Sin City*, is fidelity to their sources. The terms of that fidelity may vary widely from one adaptation and one observer to the next. Even so, filmgoers anticipating a new version of *Jane Eyre* or *X-Men* or the television audience for the BBC or *Masterpiece Theater* expect every adaptation to keep faith with the source that inspired it by following an analogous version of Wikipedia's ban on original research. Most recent adaptation theorists, however, have rejected the ideal of fidelity to the original as chimerical, impossible to achieve, and perhaps even unwise to attempt. This attack on fidelity, as several reviewers have pointed out, has become so ubiquitous in the field that it has assumed the status of a cliché itself.[51] These theorists see adaptation instead as an instance of a process whereby texts constantly generate new texts and new readings. Because every text, every interpretation, every reading is rooted in earlier texts and textual encounters, it is hard to argue that there is a categorical difference between original and adapted texts. This analogy, which places primary emphasis on textual production and originality rather than textual reproduction and adherence to past models, is surely closer to the ceaseless production, elaboration, and revision characteristic of Wikipedia than the analogy of textual fidelity.

PARADOXES OF MUST PUBLISH ORIGINAL RESEARCH

One last analogy is most illuminating of all: scholarship in the liberal arts. This model is actually an antimodel, for scholars are required to produce exactly the original research Wikipedia proscribes. The definition of "original" can vary widely from discipline to discipline and scholar to scholar, but academic scholarship depends on new discoveries, new interpretations, and new theories. What is most striking about what we might call the scholarly policy of must produce original research, or MPOR, is that it entangles researchers in paradoxes of their own.

The first of these is the problem of getting a hearing for original ideas and discoveries. MPOR requires scholarship that is genuinely original in order to escape the charges of servility and superfluity. Few scholars attract these charges in print, but many graduate students do in their teachers' comments on their coursework. En route to becoming scholars, they must overcome the *Reader's Digest* temptation to recycle summaries of their reading and instead present new theories, readings, or factual discoveries. Their failure to do so, either as graduate students or as established scholars, typically results not in attacks in scholarly quarterlies but in rejection letters from those quarterlies' editors explaining that the article they have submitted is not original enough to warrant publication.

At the same time, fledgling scholars are well advised to avoid radical theories, interpretations, and discoveries that the academic establishment is likely to reject out of hand. Great as the dangers of servility are for a budding career, the dangers of any perceived heresy or apostasy are still greater, as the careers of Galileo and Michael Bellesisles demonstrate. For every scholar capable of establishing a broadly influential new way of reading or thinking about an established discipline, there are dozens of scholars whose equally original ideas make no impact on an academic establishment whose curatorial nature resists change even as it demands originality.

This paradox surfaces most clearly in academic citations, which are designed to establish a given writer's authority by demonstrating a comprehensive knowledge of the field even as they limit that authority by acknowledging how much of what he or she has to say depends on the contributions of earlier scholars. Citations, which express both humility ("I stand on the shoulders of giants") and arrogance ("look at everything I have read"), establish a given scholar's claims to authority even as they provide cover for those claims by ascribing arguable propositions to whoever is being cited. Instead of acting like Wikipedia's citations, which serve as a guarantor of NOR, academic citations are meant simultaneously to link and to distance each original contribution to a series of ongoing debates that are presented as both vital and sadly inadequate. They amount to a sign of disavowal that serves to demonstrate both what the current monograph or essay is and what it is not.

Some of the phrases I have just used to describe this situation—"getting a hearing" and "no impact on the academic establishment" and "a sign of disavowal"—indicate a second paradox of MPOR. What exactly does it mean to get a hearing or have an impact on a given academic discipline, or (as much more seldom happens) on the academy generally? Directors of dissertations in the humanities push PhD students into highly specialized topics and questions they can plausibly master in the limited time they have to write even though the resulting narrow focus of their research is likely to have little impact on their field. By contrast, contributors who are originating, expanding, or correcting Wikipedia articles can reasonably expect to attract immediate attention from interested readers. These contributors are well aware that their contributions are likely in turn to be edited, perhaps even reverted by some of these readers. But that is not what they intend. Contributors commonly proceed as if they believed that their edits made each article complete and perfect. Their sneaking suspicion that other contributors whose views are different may well change what they have written makes them accept such changes but not welcome them. Indeed Wikipedia's many edit wars make it clear that not all contributors accept changes until they are compelled to do so. The impact of both contributions to Wikipedia and fledgling research in liberal education, in other words, is measured in terms likely to bring the contributors little satisfaction.

Academics offering new evidence for new interpretations or theories have very different expectations of what constitutes acceptance within their disciplinary communities. Since very few scholars have the power to reframe their disciplines, most scholars, even if their aims are revolutionary, settle for winning respect for their voices within a series of ongoing debates. When every scholar is surrounded and succeeded by other scholars motivated by the MPOR imperative, the most a given scholar can reasonably expect is to participate meaningfully in disciplinary debates rather than winning them. Since the goal of scholarship is to continue and renew what the Great Books Foundation calls "the great conversation,"[52] making a mark in an academic field is likely to depend more on provoking arguments than on resolving them. Fortunately for scholars, though not perhaps for the larger culture, their professional success does not depend on redefining their fields, establishing canons that stand the test of time, or advancing interpretations no one questions. Instead, professional advancement customarily depends on continued publication in what Wikipedia calls reputable sources.

On this point Wikipedia and the academy agree: it matters much less what you have to say than whether and where it is published. By framing some contributions as more prestigious, more significant, more important than others, publication venues go a long way toward establishing norms for both scholarly influence and professional success. But, unlike Wikipedia, whose goal in principle is a comprehensive compendium of human knowledge, humanistic scholarship hopes to maintain the debates that sustain it indefinitely. As Stanley Fish has observed of the labors of scholarly organizations devoted to the study of specific authors or problems, MPOR leads to a situation as fraught with paradox as the NOR imperative, for "it is the business of these societies first to create the work and second to make sure that it will never get done."[53]

Fish's remark hints at a third paradox behind MPOR: the difficulty of distinguishing what philosopher of science Thomas Kuhn has called "normal science"—that is, "research firmly based upon one or more past scientific achievements, achievements that some particular scientific community acknowledges for a time as supplying the foundation for its further practice"—and the revolutionary scientific works by Aristotle, Ptolemy, Newton, and

Lavoisier that became foundational because "their achievement was sufficiently unprecedented to attract an enduring group of adherents away from competing modes of scientific activity" and "sufficiently open-ended to leave all sorts of problems for the redefined group of practitioners to solve."[54]

Kuhn's landmark account of the sociology of scientific revolutions as based on "paradigm shifts" has been widely applied to scholarship in the humanities as well. Here, however, the distinction between what might be called normal humanities scholarship and revolutionary paradigm shifts is much harder to make. Kuhn himself acknowledges that it is no easy matter to "differentiate normal science from science in a crisis state," for "there is no such thing as research without counterinstances." But normal science, he argues, is not most aptly described as "testing or as a search for confirmation or falsification. Instead, its object is to solve a puzzle for whose very existence the validity of the [consensual] paradigm must be assumed. Failure to achieve a solution discredits only the scientist and not the theory."[55]

This distinction may well hold true for experimental science, but not for the practice of law, in which every civil and criminal action simultaneously assumes the authority of the laws relevant to the case at hand and provides potential material for challenges to those laws. Especially in cases in which lawyers on opposite sides of the aisle support their competing arguments with legal precedents that seem to contradict each other, the law as well as the defendant is put on trial. Journalists rightly label Supreme Court decisions like *Brown v. Board of Education* and *Citizens United v. Federal Election Commission* as landmark cases because they mark paradigm shifts in the law of the land, but the processes that lead to such epochal decisions are part and parcel of the normal practice of law.

Along the same lines, cultural historians can readily discern retrospective paradigm shifts like the Renaissance, the Enlightenment, and the rise of poststructuralism that are presumably to be contrasted with the normal work of the humanities that goes on in academic quarterlies and college classrooms. But normal humanities is as paradoxical a term as normal law, for it is the task of both attorneys and humanities scholars constantly to challenge the assumptions that establish their fields and enable them to work at all. Scholars who are applying consensual theoretical paradigms to the interpretation of *Hamlet* do not feel that their work is less important than that of theorists. They may be taking their theories for granted, but they are risking the opposition of whoever has been reading or teaching *Hamlet* according to the old paradigm. In fact, such reinterpreters often express impatience with theorists as abstract, remote, and unmoored from the realities of the profession. To the extent that they remain healthy, liberal education and humanities scholarship are constantly in ferment. The surest sign that they are moribund is uncontested agreement among their practitioners.

Another paradox of MPOR arises because of a conflict that has no close analogy in the case of Wikipedia. Academics are required to produce original theories, interpretations, and evidence in their role as publishing scholars. But they are forbidden to do so in their role as classroom teachers. This conflict lies at the heart of academics' professional life. Teachers of English composition, conversational Spanish, and American history may be producing cutting-edge work in their respective disciplines, but the students in these courses would be ill-served if their teachers substituted expositions of their latest research for training in the fundamentals of sentence structure, vocabulary, and historical evidence. In fact, a common complaint of students in these courses is that their teachers focus on their current research interests, which seem arbitrary and recondite, instead of the elementary material students need and expect from them. The challenge of infusing the teaching of introductory courses with the excitement of original scholarship without simply substituting the scholarship for the material proper to an introductory course is a favorite theme of anatomies of liberal education.[56] This paradox has no parallel in Wikipedia not only because contributors are prevented from sharing the fruits of their original research but because they do not see themselves as teachers, only variously expert expositors of someone else's views.

Kuhn suggests one last paradox of authority within the paradox that arises from the conflict between the demands of teaching and research. Observing "the manner in which science pedagogy entangles discussion of a theory with remarks on its exemplary applications," he notes that "given the slightest reason for doing so, the man who reads a science text can easily take the applications to be the evidence for the theory, the reasons why it ought to be believed." In truth, however, "science students accept theories on the authority of teacher and text, not because of evidence." Kuhn's point—"the applications given in texts are not there as evidence but because learning them is part of learning the paradigm at the base of current practice"[57]—is to dispel a widespread error about the source of authority for a given consensual paradigm. In liberal education, however, this argument has an important corollary Kuhn does not pursue: the conflict between what teachers and students want the students to learn. Students who see the theories, interpretations, and evidence they will be asked to regurgitate on the exam as the material of their courses reasonably ask, "Will this be on the exam?"

and seek to memorize the particulars of the *Aeneid* and *The Wealth of Nations* in order to demonstrate their mastery of the material. Their teachers, however, see these texts as examples of paradigmatic ways of thinking and their own classroom teaching as a model for students' learning rather than its proper subject. Specific examples provide the most vivid and memorable ways to present paradigms, but it is the paradigms, not the examples, students need to learn and learn to challenge. Memorizing every detail of the examples the teacher happens to have chosen misses the point of the course and also fills students with unearned confidence that they understand paradigms they are merely taking for granted.

Authority in the Academy

Even if it were not true that many contributors to print encyclopedias have been college teachers, the paradoxes of authority displayed by all encyclopedias would illuminate still further paradoxes in the academy's own attitude toward authority. The first of these is even more widely recognized in liberal education than in the compiling of encyclopedias. Although intellectual authority is established by a researcher's mastery of a given field, the greater the mastery, the narrower the field. Indeed one of the most common of all academic caricatures is the specialist who knows practically everything about practically nothing and practically nothing about everything else. This paradox is even more poignant in liberal education than in technical or professional education, because an education in the liberal arts is supposed to produce well-rounded citizens capable of thinking critically about a wide variety of subjects and issues, ideally including those in which they have never had courses. It is most poignant of all in the selection of the dissertation topic for students of the humanities, who must choose a narrowly focused topic on which no one else has published to demonstrate the expertise that will earn an advanced degree.

Liberal education seeks to produce generalists whose intellectual skills are broadly applicable and readily transferable. But few college teachers are themselves generalists, especially at research universities that have significant graduate programs. This model has cast a lengthening shadow over undergraduate education ever since what Christopher Jencks and David Riesman have called "the academic revolution" established "research, rather than teaching or service," as defining what Louis Menand dubs "the paradigm of the professor—not only in the doctoral institutions, but all the way down the institutional ladder."[58] Teachers in modern universities are expected to act as specialists in graduate seminars and as generalists in undergraduate lecture halls. More fundamentally, however, universities assume that students will get a more well-rounded education from three dozen teachers than from any one of them alone, because absorbing a range of specialized knowledge and disciplinary rules encourages students to make general intellectual and operational connections among those disciplines and their methodologies that no single teacher will make for them. The education of a single citizen requires the contribution of dozens of experts. Just as encyclopedias are compiled rather than written, liberal education arises from a confluence of specialists whose authority is greater than the sum of its parts.

But this well-known paradox is only the beginning of the subtle, nuanced, and often contradictory attitude to authority revealed by the academy's attitude toward the encyclopedias to which academics often contribute themselves. Like the administrators of Wikipedia, college teachers insist on precise and transparent citations in their students' written work. Every phrase or statistic a writer borrows from somewhere else should be cited, both in order to make it easier for readers to identify and consult their sources and in order to help the teacher distinguish the student's own contributions from the contributions a given paper cites. Within this general rule, there are further distinctions between better and worse sources to cite. Wikipedia itself, of course, is at the bottom of the list, the citation most likely to diminish rather than enhance the paper's authority. But print encyclopedias do not rank much higher. For years, high school teachers discouraged their students from citing encyclopedias, and they were generally considered beyond the pale for undergraduate and graduate writing, because they were never the best sources. Instead, encyclopedia articles were widely viewed as secondary or tertiary sources that lost the authority of the best scholarly sources without adding any new kinds of authority that would strengthen a given paper, though using them certainly made the paper easier to write.

The protocols of scholarly writing give the greatest weight to two kinds of citations: references to the most up-to-date sources that indicated a grasp of the most recent developments in a given field and references to primary sources that are seen as cutting through incrustations of distortion and interpretation to act as sources in the original sense of the term. In many instances the weight given primary sources privileges them over reprint editions, though not necessarily over revised editions incorporating the latest scholarly wisdom. In areas more clearly attuned to rapidly changing theories in the face of new discoveries, like science and technology, primary sources are of merely historical interest. Medical schools, for example, are reluctant to admit graduates of St. John's College of Annapolis, Maryland, whose Great Books curriculum limits its students' knowledge of science to

Galileo, Faraday, and Darwin and their knowledge of medicine to Hippocrates, Galen, and Freud, unless they have done considerable outside reading or gone somewhere else for a supplementary year of courses specifically designed to prepare them for medical school.

The paradoxical tension between these two kinds of sources leaves its mark everywhere in liberal education and scholarly publishing. Teachers and students debate how important it is to read classical literature in its original language. Reviewers question the desirability of historical distance from a biographer's subject, which makes it impossible to interview people who have known the subject but gives the biographer greater access to earlier biographies and multiple historical perspectives. Scholars teach their students to make sure their citations are as current as possible even as they warn them not to cite recent quotations of earlier material but to go back to the original texts and cite them directly. The only reason this tension comes into play so seldom in the academy's attitudes toward citations from encyclopedias is that encyclopedias by design never constitute either cutting-edge scholarship or primary sources. It is no surprise then that the encyclopedia entries college teachers write themselves are generally discounted and often ignored in the cases they make for promotion and tenure, for many university review committees, especially at research institutions, consider them work-for-hire rather than true scholarship. Indeed, universities and their academic units are much more likely to spell out in their policies for promotion and tenure than in their syllabuses or classrooms what sorts of authority they consider different kinds of writing to carry under what circumstances. These distinctions are often lost to students, as to users of encyclopedias generally, unless they become college teachers themselves.

Each of these paradoxes can be addressed, and some of them resolved, in practice. Teaching seminars rather than lecture courses, for example, is meant to call into question the institutional classroom hierarchy that establishes the teacher as sole authority. Inviting students to sit around a table, facing each other as well as the teacher, encourages students to talk to each other and implies that everyone at the table has equal authority to generate, challenge, and modify ideas in the course of the seminar. And seminars are famously serendipitous, often leading to investigations and conclusions far from anything the teacher had in mind. Despite the ways in which seminars share and diffuse authority, however, neither teachers nor students are ever in doubt about who is in charge and why.

All these paradoxes arise from the fact that, like Wikipedia, liberal scholarship and classroom teaching are both product and process, institution and activity, noun and verb. In some ways these paradoxes are surprisingly similar to the paradoxes behind Wikipedia's claims to authority. In others they are mirror images of Wikipedia's paradoxes, rooted in the requirement to produce original research rather than in the prohibition against it. In still others—notably, in the conflicts between academics' roles as teachers and researchers—they are neither close parallels nor mirror images. In every case, however, teaching about Wikipedia casts them in a new perspective by inviting both students and teachers to think about authority in illuminating new ways.

Despite, or perhaps because of, its potential to raise uncomfortable questions about the nature of authority in institutions that have long exercised it, Wikipedia has provoked considerable backlash, both within the academy and outside. Since criticisms of Wikipedia reveal as much about the critics' principles as they do about Wikipedia's, the following chapter examines them in detail.

CHAPTER THREE

The Case against Wikipedia

THERE ARE MANY DIFFERENT WAYS to make a case against Wikipedia. The Romanian philosopher, novelist, and aphorist Sorin Cerin, piqued by the repeated deletion of pages devoted to him in both the English- and the Romanian-language Wikipedia, has described his battles against editors who first deleted the pages on the grounds that he was not important enough. Then, when he provided evidence concerning the sales of his books and his appearance on numerous television programs, the editors refused to restore the pages because it was against Wikipedia's policy to intervene in such disputes. Cerin concludes, "Everything is based on lies and disinformation in tactics of these Wikipedians!"[1] Daniel Brandt's Wikipedia Watch objects to Wikipedia because it exerts "a massive, unearned influence on what passes for reliable information. Search engines rank their pages near the top. While Wikipedia itself does not run ads, they are the most-scraped site on the web [that is, the site whose entries are most likely to be quoted, with or without acknowledgment, by other sites]. Scrapers need content—any content will do—in order to carry ads from Google and other advertisers. This entire effect is turning Wikipedia into a generator of spam."[2] Academic historian Neil L. Waters compares Wikipedia to *Family Feud*, the quiz show in which families compete for cash awards by trying to guess, not what the right answers to a given question are, but what were the most popular answers given when program employees surveyed one hundred people. Among the winning answers to the question "Name a country with a big army" were Israel and Germany; according to the program's survey, cities about which many popular songs have been written include Texas and Oklahoma. Waters contends that "all too often, democratization of access to information is equated with the democratization of the information itself, in the sense that it is subject to a vote," and observes: "A representative of the Wikimedia Foundation (www.wikipedia.org), the board that controls Wikipedia, stated that he agreed with the position taken by the Middlebury history department [forbidding students from citing Wikipedia in their papers], noting that Wikipedia states in its guidelines that its contents are not suitable for academic citation, because Wikipedia is, like a print encyclopedia, a tertiary source."[3] James Gleick, not a particularly censorious observer, describes Wikipedia as a "scrappy, chaotic, dilettantish, amateurish, upstart free-for-all."[4]

The Quintessence of Web 2.0

Gleick's characterization is a reminder that Wikipedia represents the quintessence of Web 2.0, to use the term first coined in 1999 by Internet design consultant Darcy DiNucci and given wide currency four years later in Tim O'Reilly's first Web 2.0 Summit. DiNucci, noting the emergence of PDAs and other hand-sized hardware devices with tiny screens that challenged web designers who were accustomed to creating a single set of pages for all screens, contended that the "fragmentary" nature of Web 2.0 marked the Web as a "world of myriad, ubiquitous Internet-connected tools" and transformed it from Web 1.0, a medium to be consumed, like radio or television—"screenfuls of text and graphics"—to "a transport mechanism, the ether through which interactivity happens."[5]

O'Reilly and John Battelle, speaking in the aftermath of the 2001 bursting of the dot-com bubble, argued that many software startups had been driven out of business because they had uncritically embraced a software-driven model. This model was typified by Netscape, which counted on the dominance of its Navigator browser program to establish standards that would allow Netscape to control the terms under which Web content was made available to users. As O'Reilly later summarized their position: "Much like the 'horseless carriage' framed the automobile as an extension of the familiar, Netscape promoted a 'webtop' to replace the desktop, and planned to populate that webtop with information updates and applets pushed to the webtop by information providers who would purchase Netscape servers." Google, by contrast, was well-positioned to survive and flourish because it was conceived from the beginning "as a native web application, never sold or packaged, but delivered as a service, with customers paying, directly or indirectly, for the use of that service." Hence, "software licensing and control over APIs [Application Programming Interfaces]—the lever of power in the previous era—is irrelevant because the software never need be distributed but only performed, and also because without the ability to collect and manage the data, the software is of little use." In a sidebar entitled, "A Platform Beats an Application Every Time," O'Reilly set Microsoft's model of "a single software provider, whose massive installed base and tightly integrated operating system and APIs give control over the programming paradigm" against the "radically different business model" represented by Google, "a system without an owner, tied together by a set of protocols, open standards and agreements for cooperation."[6]

O'Reilly traced "the success of the giants born in the Web 1.0 era who have survived to lead the Web 2.0 era" to their "embrac[ing] the power of the web to harness collective intelligence." He cited Wikipedia, "an online encyclopedia based on the unlikely notion that an entry can be added by any web user, and edited by any other," as "a radical experiment in trust, applying Eric Raymond's dictum . . . that 'with enough eyeballs, all bugs are shallow,' to content creation." O'Reilly considers Wikipedia a stellar example of one of the leading design patterns that defined Web 2.0: "the key to competitive advantage in internet applications is the extent to which users add their own data to that which you provide." For O'Reilly, Wikipedia represents above all a new business model, a program for economic survival in an industry very recently decimated. It is in the interest of sites like Wikipedia that are typical of Web 2.0 to encourage users to create content because new content adds economic value.

Against Crowdsourcing

Its reliance on open sourcing, often called crowdsourcing, quickly became, for better or worse, the most distinctive feature of Wikipedia. The ability of users to add and edit Wikipedia pages at will has been the focus of most of the attacks on it. The most widespread complaint about Wikipedia is that the process of peer editing that relies on successive contributors cannot possibly winnow misconceptions and avert digital vandalism, maintain a literate and consistent style, and provide adequate documentation for each article's factual claims. Despite Andrew Lih's assumption that "a well-accepted set of general interest subjects in Wikipedia should be in good standing and reputation, because they have been heavily visited and edited by many different users on the Internet,"[7] dissenters argue that, absent a centralized team of authoritative experts and editors, Wikipedia is nothing more than an unusually unvarnished avatar of the marketplace of ideas, in which there is no evidence, only hope, that good ideas will drive out bad. Suspicion of Wikipedia on these grounds prevails throughout the academy, despite a recent survey that found that some three-quarters of both undergraduates and their teachers at Britain's Liverpool Hope College identified themselves as users of Wikipedia.[8]

The most forceful exponent of this critique is Andrew Keen, who has framed a highly influential case against what he calls the "ethical assumptions about media, culture, and technology" that underlie Web 2.0. In a 2006 *Weekly Standard* article that contained the nucleus of his more detailed but essentially identical polemic in *The Cult of the Amateur,* Keen briskly enumerates these assumptions and even more briskly dismisses them. First among them is that the interactive Web "worships the creative amateur: the self-taught filmmaker, the dorm-room musician, the unpublished writer. It suggests that everyone—even the most poorly educated and inarticulate amongst us—can and should use digital media to express and realize themselves. Web 2.0 'empowers' our creativity, it 'democratizes' media, it 'levels the playing field' between experts and amateurs. The enemy of Web 2.0 is 'elitist' traditional media."[9]

In *The Cult of the Amateur,* Keen wastes no time attacking Wikipedia, "the third most trusted site for information and current events," as "the blind leading the blind—infinite monkeys providing infinite information for infinite readers, perpetuating the cycle of misinformation and ignorance."[10] His 2006 *Weekly Standard* article already included the criticisms of Web 2.0's glorification of "narcissism," "personalization," and "the realization of the self." Traditional publishing venues employ gatekeepers, content commissioners, and filters like editors, publishers, and paying customers, to encourage the development of original works and ideas. The powers behind Web 2.0, by contrast, rely on an antiestablishment ethos that promotes intellectual parasitism, so that "instead of Mozart, Van Gogh, or Hitchcock, all we get with the Web 2.0 revolution is more of ourselves."

An even more apocalyptic assumption Keen discerns in Web 2.0 in an interview for the Ijsbrand van Veelen's 2008 Dutch television documentary *The Truth According to Wikipedia* is "that there are no truths, that everybody has their truth." The absence of gatekeepers, "the ones who determine truth," means that "anyone's truth becomes as credible as anyone else's," and "truth itself is a casualty. . . . Truth becomes truthiness," using the term television personality Stephen Colbert coined to describe unshakable convictions about the truth unmoored from logic or evidence.[11] This critique is readily applied to Wikipedia, whose cofounder Larry Sanger acknowledges in an interview in the same film that most of the first articles posted on the site "were of very poor quality, but we didn't care," because he and cofounder Jimmy Wales saw Wikipedia as "a content-generating project" rather than “a publishing project.”[12] Indeed Colbert himself, speaking on the 31 July 2006 episode of *The Colbert Report,* coined the word “wikiness” to mean “truth by consensus”: “Who is *Britannica* to tell me that George Washington had slaves? If I want to say he didn’t, that’s my right. And now, thanks to Wikipedia, it’s also a fact. We should apply these principles to all information. All we need to do is convince a majority of people that some factoid is true. . . . What we’re doing is bringing democracy to knowledge.”

Despite their reservations about Wikipedia, many intellectuals have been reluctant to go as far as Keen’s assertion in van Veelen’s film that it embodies a “cult of the amateur” in which “the less you know, the more you know.” Whether or not they agree with every particular of Keen’s critique, however, teachers who forbid their students from citing Wikipedia and encourage them not to consult it are presumably endorsing a contrary series of what Keen might call “ethical assumptions about media, culture, and technology” that underlie their own practice and the principles they wish their students to adopt. Keen’s critique in “Web 2.0” is useful not because it is precise or correct but because it implies so many contrary assumptions, principles, or ideals on whose behalf Web 2.0 in general and Wikipedia in particular have been regarded with suspicion. Since Keen identifies only a small number of these principles explicitly, it is worth teasing out the ones he does not.

Seduction versus Rectitude

The first of the ideals Keen values is implicit in the opening sentence of “Web 2.0”: “The ancients were good at resisting seduction.” Web 2.0, he warns, is one of the “great seductions,” one of the “utopian visions that promise grand political or cultural salvation.” Keen traces the roots of this seduction to “the counter-cultural utopianism of the ’60s and the techno-economic utopianism of the ’90s,” and still further back to Marx’s “seductive promise about individual self-realization” in *The German Ideology*: “whereas in communist society, where nobody has one exclusive sphere of activity but each can become accomplished in any branch he wishes, society regulates the general production and thus makes it possible for me to do one thing today and another tomorrow, to hunt in the morning, fish in the afternoon, rear cattle in the evening, criticise after dinner, just as I have a mind, without ever becoming hunter, fisherman, shepherd or critic.”[13] Dissenting from the speech at the 2005 Technology Education and Design show in which Kevin Kelly announced that “we have a moral obligation to develop technology,” Keen proposed: “Instead, let’s use technology in a way that encourages innovation, open communication, and progress, while simultaneously preserving professional standards of truth, decency, and creativity. That's our moral obligation."[14]

Keen's rejection of the narcissism, personalization, and amateurism of Web 2.0 in the name of innovation, open communication, progress, truth, decency, and creativity seems clear and ringing. But apart from the word "professional," the most important ideals behind Keen's polemic go unstated in this peroration. The ideals that would best equip defenders of culture and the arts to resist the blandishments of Web 2.0 are evidently aesthetic taste, communitarianism (although certainly not communism), and the personal rectitude that allows citizens to resist seduction—for example, the temptation to plagiarism, an accusation that was leveled against the first two volumes of the *Encyclopédie* on their publication in 1751–52.[15]

Freedom to Edit versus Freedom from Error

A second ideal implicit in Keen's defense of truth versus truthiness is freedom from factual error. This inerrancy is a value distinct from incompleteness, individual or systemic bias, arbitrariness in the selection of topics, disproportionate coverage of trivial topics, the hijacking or sanitizing of articles by associates of the public figures mentioned in those articles, stylistic obscurity or infelicity, and the often alleged cultishness of Wikipedia's inner circle or its volunteers generally—all topics that are reviewed in the Wikipedia article "Reliability of Wikipedia."[16] From its beginnings, a fundamental debate about the positive or negative value of Wikipedia has turned on the question of its freedom from the kinds of errors that can be introduced by ignorance, carelessness, vandalism, or the stubbornness of Wikipedia editors who reject changes to pages in which they are particularly invested.

The flashpoint for this debate was an article the British journal *Nature* commissioned for its December 2005 issue. Its author, Jim Giles, submitted forty-two pairs of articles on scientific topics from Wikipedia and Britannica Online, without identifying the sources of any of the articles, to a series of anonymous reviewers whom he asked to comment on their factual errors, critical omissions, and misleading statements. The reviewers found a total of 285 errors in the eighty-four articles, 162 in Wikipedia and 123 in Britannica Online: an average of four errors in each of the Wikipedia articles and three errors in each article from Britannica Online. The reviewers judged eight of these errors, four each from Wikipedia and Britannica Online, to be serious.[17] On 23 March 2006, the *Encyclopaedia Britannica* published on its website "Fatally Flawed," a report that disputed Giles's leading conclusions about *Britannica*'s inaccuracies. "Fatally Flawed" noted that Giles's reviewers had found fully a third more errors in Wikipedia than in Britannica Online. It claimed that many of the alleged errors had been found in online versions of the *Britannica Book of the Year* or *Britannica Student Encyclopaedia* rather than "our core encyclopedia." It argued that the study made no distinction between minor and major errors. It identified many more alleged errors as matters of emphasis or interpretation rather than factual inaccuracy. And it accused several anonymous reviewers, whose original reports *Nature* refused to supply to *Britannica,* of factual errors of their own.[18] Mathieu O'Neil has aptly summarized "the *Britannica* perspective" as the position that "in all cases of user-generated content—whether an Amazon review or a Wikipedia article—the *quality* of the eyes examining a project trumps their *quantity*."[19] On 23 March, in reply to *Britannica*'s demand for a retraction, *Nature* published a response that rejected most of *Britannica*'s claims.[20] The journal elaborated its position in an editorial in its issue of 30 March[21] and in a rebuttal that refused to retract the original article's comparative analysis.

Both *Britannica*'s litany of complaints and *Nature*'s rebuttal are most notable for a surprising omission. The rebuttal summarizes *Britannica*'s leading complaints as follows:

- *You reviewed text that was not even from the Encyclopaedia Britannica*
- *You accused Britannica of "omissions" on the basis of reviews of arbitrarily chosen excerpts of Britannica articles, not the articles themselves*
- *You rearranged and reedited Britannica articles*
- *You failed to distinguish minor inaccuracies from major errors*
- *Your headline contradicted the body of your article*[22]

Wikipedia evidently celebrated this incident in an entry entitled "Errors in the *Encyclopaedia Britannica* that have been corrected in Wikipedia." James Gleick cites this page in 2011,[23] but alas, it is no longer there, though some of its material has presumably been incorporated into "Reliability of Wikipedia" and associated entries.

Setting aside the relative merits of *Britannica*'s complaints and *Nature*'s rebuttals, what is most revealing here is that at no point did the *Britannica* report claim that its encyclopedia was free of factual errors. It merely claimed that those errors were less frequent in Britannica Online than in Wikipedia, less frequent in "our core encyclopedia" than in the online versions of *Britannica Book of the Year* and *Britannica Student Encyclopaedia,* and less serious than the *Nature* study contended. Indeed *Britannica* noted that, "in rebutting *Nature*'s work, we in no way mean to imply that *Britannica* is error-free; we have never made such a claim. We have a reputation not for unattainable perfection but for strong scholarship, sound judgment, and disciplined editorial review."[24] So the ideal of factual inerrancy remains in this analysis only an ideal, and the best guarantees responsible researchers can offer of its wholehearted pursuit of this ideal are scholarship, judgment, and editorial review. The assumption that *Britannica* is more likely to adhere to policies that promote this ideal is surely at least one of the reasons that in a recent experiment, "people of all ages assessed information as more credible when it appeared on the Encyclopaedia Britannica page and less credible when it appeared on Wikipedia's page."[25] Even so, this ideal is on especially shaky ground in the humanities, for, as literary critic Robert Gorham Davis pointed out a generation ago, scholarship in the humanities, as against the hard sciences, is so resistant to correcting errors in individual contributions and certifying that disciplines like history and literary studies are free from factual error that "a large proportion of what is asserted, what is taught, *cannot* be called 'true.' "[26]

Networking versus Originality

A third stricture against Wikipedia, and against Web 2.0 generally, is that it replaces creating new content by artists like Mozart, Van Gogh, and Hitchcock with networking a series of received ideas or facts through hyperlinks. The economic and intellectual value of Google, for example, is based entirely on its organization of the sites to which it leads, not on any new knowledge that it creates. The contrasting ideal is clearly originality, the ability to create new ideas, experiences, and cultural possibilities.

Despite the obvious appeal of a contrast between creation and networking, it is hard to draw a bright line between originality and compilation. Even the most original-seeming artistic and intellectual ideas come from somewhere. Indeed if they are perceived as too original, they often fail to find an audience. The eighteenth-century Scottish philosopher David Hume's rooting of creativity in the ability to recall earlier associations and arrange them in new ways strikes at the heart of this distinction even as it eerily prefigures the architecture of Web 2.0. Contemporary research in brain chemistry offers little support for the argument that creativity is a distinctive mental operation.[27] As Jonathan Lethem observes in "The Ecstasy of Influence":

> The kernel, the soul—let us go further and say the substance, the bulk, the actual and valuable material of all human utterances—is plagiarism. For

> substantially all ideas are secondhand, consciously and unconsciously drawn from a million outside sources, and daily used by the garnerer with a pride and satisfaction born of the superstition that he originated them; whereas there is not a rag of originality about them anywhere except the little discoloration they get from his mental and moral caliber and his temperament, and which is revealed in characteristics of phrasing. Old and new make the warp and woof of every moment. There is no thread that is not a twist of these two strands. By necessity, by proclivity, and by delight, we all quote.[28]

Except that Lethem didn't make this observation himself. Like so much else in his tour de force essay, not only the sentiments but the very sentences are cribbed from elsewhere—in this case, the first two from a letter Mark Twain wrote to console Helen Keller, who had been accused of plagiarism,[29] the last three from Ralph Waldo Emerson's "Quotation and Originality."[30]

It is particularly difficult to isolate originality within highly formulaic genres like the whodunit, the Hollywood Western, and the academic essay. Is a new approach to a familiar subject truly original? Is a list of topics or qualities? Is a list of examples that illustrate or complicate those qualities? In practice, academic writing seems to combine a demand for originality with a proscription against too much originality or originality of the wrong sort. In at least one context Keen does not consider, however, originality compels respect: the continuing academic proscription against plagiarism, which is still widely demonized as the cardinal sin on college campuses. In schools like the College of William and Mary, Princeton University, the University of Virginia, Haverford College, Davidson College, and the United States service academies, students are required to sign a statement indicating that they have neither given nor received outside assistance on a given assignment.

Struck by the rise of Web 2.0 and related technologies—and curious about the ways they might illuminate the value liberal education has traditionally accorded originality—Kenneth Goldsmith began in 2004 to offer an undergraduate course at the University of Pennsylvania on "Uncreative Writing." His course description began:

> It's clear that long-cherished notions of creativity are under attack, eroded by file-sharing, media culture, widespread sampling, and digital replication. How does writing respond to this new environment? This workshop will rise to that challenge by employing strategies of appropriation, replication, plagiarism, piracy, sampling, plundering, as compositional methods.[31]

The assignments Goldsmith gave his students—type out five pages of text copied verbatim from another source, transcribe an audio recording of a speech or interview, scrawl retro graffiti in public places, prepare an after-the-fact screenplay from a film like a home video or porn feature that was not based on a screenplay—redirected the attention they would normally have given the novel aspects of these texts to their materiality, nuance, performativity, and recycling of familiar tropes. When Goldsmith invited a visiting lecturer to his class and instructed his students, unbeknownst to the speaker, to share their impressions through the class listserv while he was still speaking, "students hypertext[ed] off the ideas of the instructor and their classmates in a digital frenzy. . . . The top-down model had collapsed, leveled with a broad, horizontal student-driven initiative, one where the professor and visiting lecturer were reduced to bystanders on the sidelines."[32]

Bias versus Neutrality

A fourth value implicit in Keen's attack on Web 2.0 as fueled by "countercultural radicals," "radical communitarians," "intellectual property communists," and "economic cornucopians,"[33] is neutrality. The phrases cited above make it clear that Keen is anything but a neutral observer himself. But he expects encyclopedias to be neutral in the sense of remaining free from obvious programmatic bias. His argument has been amplified by conservative commentators who find in Wikipedia a systemic bias in favor of political liberalism. One of these commentators, attorney Andrew Schlafly, the son of antifeminist activist Phyllis Schlafly, has created Conservapedia as a corrective to Wikipedia. Dissatisfied with Wikipedia's description of the U.S. Democratic Party—"In recent decades, the party has adopted a centrist economic and socially progressive agenda, with the voter base having shifted considerably. Today, Democrats advocate more social freedoms, affirmative action, balanced budget, and a market economy tempered by government intervention (mixed economy)"[34]—Conservapedia offers an alternative represented by its section on Foreign and Military Policy, which reads in toto:

> According to its platform, the Democratic Party has the objective of strengthening America. Democratic national leadership has been accused of being ambivalent about terrorism and insufficiently patriotic. A poll conducted by Fox News released in October 2007 found that 1 in 5 Democrats—nearly 10 million voters—think the world will be better off if the United States were to lose the War in Iraq. The poll found this sentiment 3 to 4 times higher among Democrats than among moderate, centrist, and Republican voters.[35]

Because it takes into account two profoundly different voices in its assessment of the Democratic Party, this summary approximates the neutral point of view, or NPOV principle (see chapter 2) as it is stated by Wikipedia: "Editing from a neutral point of view (NPOV) means representing fairly, proportionately, and, as far as possible, without bias, all of the significant views that have been published by reliable sources on a topic."[36] Readers who take the inclusion of both voices as a sign of neutrality, however, will find Conservapedia's entry for the Republican Party remarkably free of any evidence of plural voices, especially negative characterizations that liberals might endorse. The closest approach to any such assessment is the entry's closing sentences assessing the strength of the party in 2009: "The GOP weaknesses were glaring: the June poll found that the Republican Party is viewed favorably by only 28% of Americans, the lowest rating ever in a New York Times/CBS News poll. In contrast, 57% said that they had a favorable view of the Democratic Party. However, it should be noted that this poll was conducted by the mainstream media and thus is a clear example of liberal bias."[37]

When so many sources of information base their claim to authority not on neutrality but on forthrightly countering what they see as the prevailing bias of other authorities, neutrality can seem impossible either to achieve or to define. Wikipedia editors have acknowledged that producing an entry on the Israeli-Palestinian conflict that would satisfy both parties often seems as difficult as resolving the conflict itself. If Wikipedia's definition of a NPOV is quite specific, however, so is the kind of neutrality favored by the academy. Neutrality is a desideratum of academic writing in only the limited sense of disinterestedness or freedom from the kinds of bias that would prejudice or disable critical judgment. And recent attacks on American colleges as bastions of uncritical liberal acculturation that wish to "dismantle the traditional curriculum" in the name of modish ideologies and provide "an education in closed-mindedness and intolerance" accuse academics of failing to achieve even that limited sense of neutrality.[38] Undergraduate essays, like the academic articles on which they are distantly modeled, are by their nature argumentative. The neutrality they inculcate, like that of a debating team, is the ability to make the best possible case for or against a given point of view. Yet this kind of rhetorical training is only preliminary to the ultimate aims of liberal education, whose alumni presumably believe what they say.

Anonymity versus Authorship

A fifth value Keen does not mention, though it is implicit in his argument, is authorship. This value is made more explicit in Randall Stross's identification of the "single nagging epistemological question" posed by Wikipedia: "Can an article be judged as credible without knowing its author?"[39] The assumption that it cannot is behind Howard Rheingold's advice about how to improve one's "crap detection" skills: "Think skeptically, look for an author, and then see what others say about the author."[40] It is not necessary for an author to be a household name, or even to be recognized within a given field, to exemplify this value, for the ideal of authorship threatened by Wikipedia concerns accountability rather than identifiability. The articles in Wikipedia are better thought of as group authored than as anonymous. Jimmy Wales, noting that every change to Wikipedia is signed by some username, proposes "pseduonymity" as occupying a middle ground between "anonymity" and "real names": "You've got a pseudonym, but it's a stable identity, and . . . you're willing to stand behind it."[41] And P. D. Magnus calls Stross's objection "a red herring" because "stories in the *New York Times* typically carry bylines, but our believing what they say does not typically depend on what we know about the specific reporter credited. The article has the authority of something printed in the *Times*. Knowing who wrote it does not usually matter. So, too, for *Wikipedia* articles."[42] But for Bob McHenry, the former editor-in-chief of *Encylopaedia Britannica* interviewed in *The Truth According to Wikipedia,* the absence of any identifiable authors fatally undermines Wikipedia's claims to authority. Instead of accepting the invitation to join other anonymous users in editing Wikipedia, he argues: "It cannot be on me to correct the errors I find in Wikipedia. The responsibility for those errors lies with the publisher. It can't be anywhere else."[43] In rejecting the possibility of peer editing, McHenry assumes a model of authorship that concentrates authority in an identifiable, and ideally in a single, authorial agent, even if that agent happens to be an editor or publisher or another of Keen's gatekeepers instead of an author.

It is no great stretch to conflate authority with authorship, since both words, etymologically so similar, come from the Latin *auctor,* derived in turn from *augeo,* to grow. An author was for the Romans someone who caused something to increase—a creator or originator, a backer or supporter. Even though the report "Fatally Flawed" took pains to distance itself from what it deemed the "Internet database that allows anyone, regardless of knowledge or qualifications, to write and edit articles on any subject," the essay itself appeared without any byline except for "Encyclopaedia Britannica Inc."[44] This byline finessed the question of authorship by proposing the publisher, on McHenry's model, as a collective author more authoritative than any individual author could possibly be. The point is that the author need not be famous or even individually identifiable as long as he or she or it is accountable to critics and readers. It is no wonder then that, when Britannica president Jorge Cauz announced in 2009 that *Britannica Online* would invite user edits and additions to its entries, all of them to be kept distinct from "Britannica checked" content, he indicated that prospective editors would need to register under their real names and addresses.

Amateurism versus Professionalism

A criticism more often leveled at Wikipedia than at Web 2.0 generally is its disorganization. It is not that articles in Wikipedia are hard to find—the site's search engine and hotlinks make them much easier to find than articles in print encyclopedias—but that individual articles, growing out of the whims of myriad individual contributors, are often disorganized, and the body of knowledge represented by the whole endeavor disproportionately weighted toward topics of current interest (Barack Obama citizenship conspiracy theories, *Dungeons and Dragons* controversies), topics of dubious staying power (Britney Spears, Harry Potter), or outright myths (Loch Ness Monster, unidentified flying objects) to which reputable encyclopedias would never give space. The relative length of Wikipedia articles is largely arbitrary, dictated by the interests of enthusiasts who post to the site or debates among self-styled experts. Johnny Hendren, blogging for Something Awful in 2007, coined the term "Wikigroaning" for a game that asked players to guess which of a pair of Wikipedia articles (prime number or Optimus Prime, arachnids or Spider-Man, Aristotle or Oprah, God or Kevin Smith) was longer, more professionally edited, and more generally perceptive.[45] The results for many of these pairs were indeed surprisingly worthy of groans.

Surely liberal education would seem to be far better organized and more proportionate in its emphases than Wikipedia. But we have come a long way from the seven arts—grammar, logic, rhetoric, arithmetic, geometry, music, astronomy—Socrates proposed as the basis for the study of philosophy. With rare exceptions like St. John's College, institutions of higher education today serve buffets of knowledge rather than offering prix fixe menus. Many new courses on topics of contemporary interest have come under sharp questioning for both their intrinsic value and their lack of integration with the rest of the curriculum. Even set curricula are increasingly taught by a rotating roster of instructors who rarely compare notes about their teaching. The result, widely lampooned by reformers, is a cafeteria approach to education hamstrung between appeals to the canon (or more likely the canon du jour) and cries for relevance. This debate has only been exacerbated by the latest turn in the corporate university, which, treating students as customers who "increasingly see their studies as an investment in their financial future,"[46] pitches courses at their current interests or their avowed economic goals rather than aiming to develop the skills and interests someone thinks they ought to have. These developments may be salutary—it may well be better to teach the conflicts, for example, than to dispense categorical but outdated information—but they are not advertisements for the relatively higher organization of liberal education. If the academy is well-organized intellectually, the organization must involve some more specific value or values.

Keen's forthright attack on the amateurism of Web 2.0 invokes one obvious countervalue: professionalism. When Keen recounts the struggles of climate modeler William Connelley, an expert on global warming at the British Antarctic Survey in Cambridge, against first "a particularly aggressive Wikipedia editor" and then "the Wikipedia arbitration committee" over his contributions to the Wikipedia entry on global warming, it is easy to understand his outrage that the arbitrators "treat[ed] Connelley, an international expert on global warming, with the same level of deference and credibility as his anonymous foe—who, for all anyone knew, could have been a penguin in the pay of ExxonMobil." When Keen focuses on the deleterious effects of the cult of the amateur on "culture and the arts" rather than the experimental sciences, however, the tokens of professionalism he would endorse are not nearly as clear. He contends that "what defines 'the very best minds' available, whether they are cultural critics or scientific experts, is their ability to go beyond the 'wisdom' of the crowd and mainstream public opinion and bestow on us the benefits of their hard-earned knowledge." But this formulation is more notable for the energy of its scare-quoted rejections than for the specificity of its positive criteria, which seem reducible to "hard-earned knowledge."[47] In *The Truth According to Wikipedia* Keen says, "I don't believe in genius," but calls for a meritocracy based on "hard work."[48] This sounds like a recipe for precisely the kind of amateurism that produced Mozart, who began composing at six; Van Gogh, who never attended art school; and Hitchcock, who after entering the film industry as an art director and title designer rapidly rose to directing pictures because he had assimilated so much knowledge about how to do so from observing superiors whom he was not afraid to challenge.

If professionals, to take a simple economic model, are people who make money from their labors, then Mozart and Hitchcock qualify, but not Van Gogh. Neither do many other amateurs: poets like John Keats, who trained as an apothecary and physician, T. S. Eliot, who worked at a bank, and Wallace Stevens, who worked at an insurance company; composers like Robert Schumann, who left the study of law only to fail in his chosen career as a concert pianist, Alexander Borodin, an engineer, and Nikolai Rimsky-Korsakov, a naval officer; or artists like Grandma Moses, a widow who taught herself to paint when she was approaching eighty.

A meritocracy based on hard work might alternatively be defined in terms of apprenticeship rather than either professional study or financial success. Apprenticeship was the unmarked model for professional credentialing in virtually every trade and vocation but the ministry until the second half of the twentieth century, when a college degree became identified as the all-purpose professional credential. There is ample historical precedent for this shift—for example, in the shift from the seven liberal arts Socrates prescribed as the basis for the (presumably amateur) study of philosophy to the trivium and quadrivium prescribed by medieval universities as a preparation for the professional study of law, medicine, or divinity. The medieval university was ultimately "a scholastic Guild whether of Masters or Students" not authorized by "King, Pope, Prince, or Prelate."[49] Its teaching and credentialing authority thus derived from its guild status, not from any outside authority figure.

American colleges are often assumed to operate under a guild model of apprenticeship. But the reign of this model was short. Before World War II, colleges were essentially finishing schools for the sons of privilege; since the 1960s they have been cast increasingly as preprofessional schools providing ports of entry to the modern multiversity. Disinterested liberal education as John Henry Newman imagined it is a nostalgic conceit rather than a real-world alternative or even an autobiographical memory for contemporary professors. The avowed goal of most college students today is preprofessional training or professional credentialing, even if they have no idea what their profession is likely to be. To what extent should a college education be preprofessional? What is the best model of preprofessional education—the rounds of medical school? the lecture classes of law school? the seminars of graduate school? And which authority—the faculty? the trustees? the students? the state legislatures and funding agencies on which so many colleges rely for funding?—ought to be making the rules? Contemporary liberal education's continued debates over these questions have mirrored rather than resolved Keen's lack of precision in defining professionalism.

One token of professionalism Keen and the academy unite in recognizing is expertise, which is distinct from both study and accreditation. But the whole project of an encyclopedia written and compiled by experts depends on a paradox he does not acknowledge, even though it had been set forth by Alexander Blair in an unsigned essay as early as 1824: "All attempts at bringing knowledge into *encyclopedic* forms seem to include an essential fallacy. Knowledge is advanced by individual minds wholly devoting themselves to their own part of inquiry. But this is a process of separation, not of combination."[50] The ideals of any encyclopedia will necessarily be at odds with the ideals of expertise.

The philosopher Stephen Turner has revealed an equally knotty paradox in the reception of expert knowledge that is rooted in the very nature of expertise:

> The basis on which experts believe in the facts or validity of knowledge claims of other experts of the same type they believe in is different from the basis on which non-experts believe in the experts. The facts of nuclear physics, for example, are "facts," in any real sense . . . only to those who are technically trained in such a way as to recognize the facts as facts, and do something with them. The non-expert is not trained in such a way as to make much sense of them: accepting the predigested views of physicists as authoritative is pretty much all that even the most sophisticated untrained reader can do.[51]

Experts can explain themselves convincingly, and with a serious possibility that their claims might be intelligently disputed, only to other experts; most of their listeners and readers will be persuaded (or not) by their status as experts, not by the claims they are making. Once their expertise is stipulated, particular experts have in a fundamental sense ceased talking to the nonexperts who most rely on their expertise.

In addition, it is clear that many amateurs have made themselves experts in their fields. The writers behind the Wikipedia entries on a wide range of role-playing and video games are obviously experts, and their expertise obviously derives from an amateur's passion, since there is no way it could have been professionally accredited. Keen's fear that Web 2.0 "levels the playing field" assumes that everyone, from professional climatologists to anonymous amateurs, will be seen as equally qualified to weigh in on a given subject. But it seems more likely that Wikipedia represents another kind of democratization, a leveling of different areas of expertise, so that it is just as worthwhile for contributors to spend hours toiling over the entry on Gandalf as the entry on nuclear physics. In this brave new world, no one is an expert on everything, but everyone is potentially an expert on something. No area of expertise is assumed to be more important than any other.

To dissent from this view, as Keen clearly does, requires distinctions between more and less worthwhile areas of expertise. Keen himself draws a sharp distinction between the established culture industries and the cult of the amateur. In doing so, however, he overlooks the fluidity of cultural capital, which allows him to give Hitchcock as an example of an established artist like Mozart and Van Gogh, even though half a century ago his critics were fighting to rescue him from the status of mere entertainer, and half a century before that motion pictures were considered a disreputable way to make a living. Yesterday's video-game enthusiast may well become today's video-game designer and tomorrow's video-game executive. The distinctions between established areas of expertise and mere cults, which seem so clear from moment to moment, are constantly subject to change.

No institution is more sensitive to both the enduring value of established areas of expertise and their incessant challenge by emerging areas than the academy, which aims to use the wisdom of yesterday to prepare the citizens of tomorrow. The culture wars of the past twenty-five years may have shaken individual departments and universities, but they may also have been tonic for liberal education as a whole. Certainly the questions they have raised have been central to the whole enterprise. In their sensitivity to both the contingency and the universality of their essential texts and their foundational questions, colleges have continued to incorporate both the democratic impetus behind Wikipedia and Keen's counterpolemical impulse and have sought to rise above both of them.

Anti-institutionalization versus Institutionalization

Keen's emphasis on "established artists" implies still another value he wishes to defend: institutionalization. Here the academy occupies a paradoxical place. It is widely regarded as an institution whose corporatization in recent years has institutionalized it still further, one that is deeply invested in a conservative view of institutional culture yet at the same time celebrated, or condemned, for fostering antiestablishment, anti-institutional ways of thinking. Hence the American college is an anti-institutional institution whose economic survival depends on persuading a critical mass of shareholders to support its conservatory mission but whose intellectual vitality depends on constantly challenging the conventional wisdom it dispenses and maintaining an environment that encourages its students to do the same.

Mutability versus Stability

Unlike many other critics of Wikipedia, Keen does not emphasize its status as a work-in-progress. But attacks on Wikipedia's mutability are legion. The instability of Wikipedia's online format, which allows even the most extensive edits to be incorporated almost instantaneously into a given article until they are in turn edited or reverted by later users, means that the reliability of a given article may depend on the date, or even the time of day, when it is consulted. Hence Wikipedia is for better or worse a living document, perhaps even, as some of its critics charge, a nondocument. Its entries have fallen victim to countless instances of transcription errors, digital graffiti, and vandalism. (My own personal favorite, long since reverted: shortly following the assassination of Osama bin Laden, my wife consulted the Wikipedia page on bin Laden and discovered, according to a recent addition, that the youngest of his children was Harriet Meirs, whom George W. Bush had unsuccessfully nominated to the Supreme Court.)

The most famous case of vandalism in Wikipedia's history to date concerns the entry of John Seigenthaler, a veteran journalist who had served in the Kennedy administration. For four months in 2005, his Wikipedia entry included the following passage:

> John Seigenthaler Sr. was the assistant to Attorney General Robert Kennedy in the early 1960's. For a brief time, he was thought to have been directly involved in the Kennedy assassinations of both John, and his brother, Bobby. Nothing was ever proven.[52]

When Seigenthaler became aware of the error, he attempted to track down the anonymous contributor who had posted this libelous statement, but BellSouth, the contributor's Internet provider, refused to help him. Nor, under the 1996 Communications Decency Act, could Wikipedia or any other online service provider, unlike print publishers and broadcast corporations, be held legally responsible for any defamatory material for which it provided space. Daniel Brandt, a longtime critic of Wikipedia's decision to value its users' privacy over their accountability, eventually succeeded in identifying the malicious poster as Brian Chase, who worked for a delivery firm in Nashville, Tennessee, and shared his identity with his victim. Once he had been provided with this information, Seigenthaler not only declined to sue Chase for damages but asked Chase's employer, Rush Delivery, not to accept his resignation. In response to Seigenthaler's revelations, Wikipedia introduced a new policy of "semi-protection" designed to discourage "drive-by vandals." Unlike Wikipedia's existing protection policy, which quarantined a small number of popular and hotly disputed articles from any editorial changes whatsoever, semi-protection, which "could be applied to any article," prevented "unregistered and newly registered users (less than four days old [that is, anointed as registered users less than four days earlier] and having made fewer then ten edits)" from editing articles designated as semi-protected.[53]

The policy struck a new balance between privacy and accountability but did nothing to address Wikipedia's reputation for instability. Indeed it emphasized the contingency of both Wikipedia itself and of its editorial policies. Bob McHenry, who dismissed Wikipedia as "the encyclopedia game, played online," when he was interviewed for *The Truth According to Wikipedia,* elaborated: "They have no mechanism for assuring, say, maximum accuracy, maximum reliability, before they publish."[54] In McHenry's view, the contingency characteristic of rough drafts should be replaced upon publication by a stability that has been earned by expert writing and editing. Wikipedia, in this view, is nothing but an endless series of rough drafts that are incessantly and publicly edited by the very community that has been consulting articles and relying on their authority. McHenry characterized the "unspecified quasi-Darwinian process [that] will assure that those writings and editings by contributors of greatest expertise will survive; articles will reach a steady state that corresponds to the highest degree of accuracy" as "entirely faith-based."[55] Against this faith-based model of evolution toward perfection, McHenry sets *Britannica*'s model of commissioned experts and editors whose work requires less later revision because it is more likely to be correct when it is first published.

Britannica itself, as McHenry acknowledges, is neither error free nor stable. That is why it has gone through fifteen print editions over the past 250 years. Although *Britannica*'s imposing row of hard-copy volumes makes it seem much more stable than Wikipedia, its articles have been revised more and more frequently over its history. Before it ceased hardcover publication in 2010, its publisher had committed since 1936 to revising every article twice every decade and has produced since 1938 an annual *Book of the Year*. Newer editions update older editions by indicating, for example, that formerly living subjects have now died and by correcting errors in earlier editions, for example, by adding Pluto to the list of planets in the fourteenth edition and demoting it to subplanetary status in its current entry in Britannica Online.[56] The fact that Pluto is still listed as a planet in the overwhelming majority of hard copies of *Britannica* sitting on library shelves indicates that the values of stability and continuity are always in conflict with other important values. *Britannica* might fairly be described as a work-in-progress that masks its contingency in order to look as reassuringly solid, stable, and immutable as possible, so that even successive *Books of the Year,* bound to emphasize their uniformity with each other and the encyclopedia itself, seem to supplement the latest edition rather than correcting it.

Britannica's history vividly dramatizes the paradoxes involved in updating a reference. Its need to present itself as both authoritative and timely means that new editions were periodically planned and constantly in progress. Its *Books of the Year,* vigorously marketed as updates that made it the last word in reference sources, were more like almanacs than encyclopedias, emphasizing as they did exactly the kinds of information (scientific breakthroughs, political developments, necrologies) most likely to be of purely current interest. As the years passed, subscribers to these updates would find their indefinitely extended shelves of *Books of the Year* increasingly useless as references because of their organization as annuals. The ideal purchaser of *Britannica,* at least from the publisher's point of view, would purchase each *Book of the Year* and then each new edition in acknowledgment that the annual updates had still failed to keep the encyclopedia up to date as a convenient reference. Few individuals subscribed on this basis, but many libraries did, reinforcing their status as reference authorities. Yet when libraries donated their outdated editions of *Britannica* to used-book sales, these sets, unlike the *Books of the Year,* were invariably among the items most eagerly snapped up by purchasers who presumably found them valuable references despite their age.

Since human knowledge and understanding are constantly changing, institutions and disciplines that use their stability to bolster their claims to authority must find ways of managing the inevitability of their own change. The institution that puts its greatest faith in stability is the Ministry of Truth in George Orwell's *1984,* in which each new version of the historical record simply makes earlier, potentially inconsistent versions disappear down the memory hole. Perhaps the closest any real-world institution comes to Orwell's vision is the College of Cardinals, in which, as church historian John Boswell was fond of saying, "You can hold a minority opinion at 2:45 and be just plain wrong by 3:15." In its discussion of whether papal pronouncements ex cathedra were to be considered infallible, the Catholic Church differed from the Ministry of Truth by preserving the particulars of the debate, but once the matter was settled by a vote, it was not to be revisited.

Other institutions manage the claims of stability and responsiveness to contemporary understanding in apparently less absolute terms that still reveal the conflicts between the two. In Thomas Kuhn's influential account of scientific revolutions, periods of consensual stability in the history of experimental science are punctuated by periods of more or less radical interrogation of science's enabling assumptions driven by new evidence or new discoveries—for example, of the Higgs boson particle—that either do or do not succeed in changing the community's thinking. If they do not, they are read out of the discipline; if they do, they become working articles of faith until they are challenged by still later discoveries. Something of this same process operates in case law. In Great Britain, case law supplements parliamentary sovereignty as the basis of legal authority. In the United States, laws are routinely reviewed by courts all the way up to the Supreme Court, which has ruled many laws unconstitutional and reversed its own earlier rulings hundreds of time. Recent American debates between Originalists who claim that any given passage of the Constitution admits of only the meaning its framers intended and Living Constitutionalists who contend that the Constitution was intended as a dynamic document subject to legal reinterpretation are only the latest chapter in on ongoing and unavoidable debate between the virtues of stability and flexibility.

Academic scholarship in the liberal arts draws in different ways on all these models. In its search to remain at once rooted in the past, responsive to the present, and informed by its own history, it emphasizes the commonalities rather than the differences among case law, ecclesiastical law, and scientific understanding. The motto of each discipline's historical understanding might be the epigram attributed to humorist Ashleigh Brilliant: "My opinions may have changed, but not the fact that I am right." Surveying liberal education's deep commitment to "freedom of opinion [that] requires a diversity of contradictory opinions," Robert Gorham Davis concludes: "In the universities as outside them, there is democracy, pluralism, and partisanship in the realm of ideas. This is only possible because ideas are not true."[57]

Inclusiveness versus Selectivity

Yet another value Keen imputes to traditional culture is selectivity. Although thousands of items are cut from Wikipedia articles every hour of every day, the general tendency, hailed by Jimmy Wales and other champions of Wikipedia, is to publish more and longer articles, and Wikipedia's stubs, its shortest articles, typically carry invitations to expand them. The result, according to skeptics, is that Wikipedia, never planned by any central authority, grows ever longer and more ungovernable.

It is true, as commentators on Wikigroaning have often observed, that the often ludicrous disproportion of trivial Wikipedia articles does not undermine the authority of more substantive articles because there is no shortage of bytes in the Wikipedia universe. Unlike *Britannica,* Wikipedia can allow any number of articles to grow unchecked without having to cut other articles to conserve space or paper or production budgets. The real point, however, is Keen's implication that the monuments of traditional culture are valuable precisely because they have been selected from among many other contenders that did not make the cut. Keen couches his defense in frankly elitist terms when he writes in "Web 2.0" that "the purpose of our media and culture industries—beyond the obvious need to make money and entertain people—is to discover, nurture, and reward elite talent." Dissenters from this view like T. W. Adorno, who see the culture industry as driven by economic imperatives and the need to maintain the political status quo, might cavil about Keen's shunting of these goals into a parenthetical clause. But these criticisms would probably not deter Keen, who adds: "Without an elite mainstream media, we will lose our memory for things learnt, read, experienced, or heard. The cultural consequences of this are dire, requiring the authoritative voice of at least an Allan Bloom, if not an Oswald Spengler."[58]

In America, the culture wars of the past quarter century have been fought precisely over the question of how selective culture should be and on what grounds. Which interests of present and past shareholders do cultures have the greatest responsibility to represent? Who has, and who should have, the authority to select which cultural artifacts will have the greatest currency today and endure tomorrow? Can a culture define itself without making invidious exclusions? In what sense can a nonselective culture truly be called a culture? In recent years these questions—first framed as attacks on a liberal educational establishment increasingly seen as willing to jettison its cultural elitism in favor of capitulation to a mob culture—have become a defining subject of liberal education itself. Colleges under assault for eliminating foreign-language requirements and replacing courses in canonical literary figures with workshops in television, video games, and Web design have responded by "teaching the conflicts," seeking to revitalize their curricula and their educational mission by reexamining their foundational assumptions more critically and comprehensively.

Producers versus Consumers

When they conduct self-examinations or make institutional changes instigated by their students and critics, colleges risk running afoul of still another of Keen's strictures: the need to maintain a sharp distinction between purveyors and consumers of knowledge. In "Web 2.0," Keen warns that "one of the unintended consequences of the Web 2.0 future may well be that everyone is an author, while there is no longer any audience." The hierarchy between authors and audiences Keen sees as threatened by the egalitarian, express-yourself impetus of Web 2.0 is mirrored in the equally threatened hierarchy between teachers and students who, as purchasers of an extremely expensive product, feel increasingly entitled to demand courses relevant to their immediate interests and career plans, seek exemptions from academic requirements, and protest grades that might hurt their chances in the job market. Universities eager to improve their rankings in the annual *U.S. News and World Report* listings seek to inflate their applicant pools and mollify matriculated students who might otherwise rock the boat are altering the balance of power between teachers and students throughout the system of higher education.

Facility versus Depth

These threats to a system once regarded with pride as selective and elitist go hand in hand with one final criticism of Wikipedia: its facility is widely taken as evidence of a fatal lack of depth. Wikipedia is seductively easy to add to, correct, and edit. It is likewise so easy to consult—no trips to the library, no paging through multivolume tomes, no chasing from one cross-reference to another—that it subverts the normal progress of research. Hence Jaron Lanier argues that Wikipedia provides not only students but "search engines with a way to be lazy," increasingly listing it as the first entry, sometimes, as on mobile devices that include "text-entry boxes and software widgets that are devoted exclusively to Wikipedia," to the virtual exclusion of other sources by identifiable authors.[59] In the process, Wikipedia has become a one-stop shopping source for many novice researchers for whom it marks both the beginning and the end of any search for information, tyrannically short-circuiting the possibility that the people who consult it will ever test its assertions against that of any other reference.

Nicholas Carr broadens and deepens this critique when he argues that reading hypertext generally, and working with Web 2.0 in particular, increases such "primitive" and "lower-level skills" as "hand-eye coordination, reflex response, and the processing of visual cues," along perhaps with "fast-paced problem-solving" and "quickly distinguishing among competing informational cues, analyzing their salient characteristics, and judging . . . [their] practical benefit."[60] Sharpening these primitive skills, however, is a small return for the toll Web 2.0 takes on its users. The distractions of hypertext and information overload make it harder to concentrate on a screen full of words without clicking through the omnipresent hyperlinks. Multimedia technologies threaten to limit rather than enhance readers' ability to acquire and retain information by straining their memories past a useful point. Veteran users of the Web tend to skim, graze, or "power-browse" sites instead of reading them through.[61] In support of his premise that the Web may be making its users dumber, Carr cites researcher

Erping Zhu: "Reading and comprehension require establishing relationships between concepts, drawing inferences, activating prior knowledge, and synthesizing main ideas. Disorientation or cognitive overload may thus interfere with cognitive activities of reading and comprehension."[62]

If it sounds too obvious to observe that the ability to read, digest, and remember written material is essential to liberal education, consider why the kind of deep thinking Carr's title, *The Shallows,* contrasts with online browsing is equally essential. Deep thinking is concentrated, sustained, reflective, self-aware, perhaps difficult to follow, even obscure. All these features are surely characteristic of the kind of thinking liberal education seeks to promote. But it also wishes to train minds that are quick, agile, self-critical, and resourceful enough to think outside the box. Carr's uncritical emphasis on reading comprehension as the indispensable preparation for deep thinking reflects the continuing attempts of my own field of English studies to define its goals in terms of the mastery of literature, a series of canonical texts to be read and comprehended, instead of the mastery of literacy, the ability to do things with the texts one has read or skimmed or grazed or power-browsed.

Keen recoils from Kevin Kelly's celebration of "Liquid Versions" of books that, once digitalized, are freely available to be "cross-linked, clustered, cited, extracted, indexed, analyzed, annotated, remixed, reassembled, and woven deeper into the culture than ever before."[63] Keen's retort—"A finished book is not a box of Legos, to be recombined and reconstructed at whim"[64]—overlooks the ways in which books are already routinely treated as boxes of Legos by readers who combine their memories of different books to produce new fantasies and ideas and researchers like Keen himself whose footnotes trace the recombinant process according to hallowed scholarly conventions. Keen's focus on the book as object, the exclusive property of its author, leads him to overlook the many ways that books are used as resources by the readers for whom they are written. The more critical relationship between literature and literacy that a closer consideration of Web 2.0 could foster seems essential to the revitalizing, perhaps to the survival, of English studies and liberal education generally.

Coda: *Britannica* versus Liberal Education

Critics of Wikipedia commonly assume that the values they see as under attack by its spread and influence are straightforward, consensual, and unproblematic. But generating a list of explicit and implicit values imperiled by Web 2.0 reveals a surprising lack of congruence between these values and the values of liberal education today. The academy has not set itself against any of the values implied by Keen's critique of Wikipedia. And it has endorsed a number of them—aesthetic taste, authorship and accountability, originality, professional expertise, deep thinking, perhaps even personal rectitude—more or less explicitly. But its relation to most of them is far more equivocal and has become ever more deeply equivocal since the Renaissance. Liberal education embraces neutrality and freedom from error in principle while tolerating widely divergent and often logically contradictory views in practice. The value the academy attaches to selectivity and elitism, to institutionalization and stability, is complicated in every case by its openness to self-criticism and the contradictory impulses it incorporates within itself. Liberal education shuns absolutism. It is most comfortable with nonabsolutist values like taste, expertise, accountability, originality, and critical thinking. But to the extent that these values, or any others, threaten to become absolute, they become suspect. The editors of *Encyclopaedia Britannica* might like to believe the values of traditional references and liberal education are congruent. But they have actually been very different ever since schools began taking their models from Plato and Aristotle—and especially from the subversive method of Socrates—rather than from the philosophers before them.

Critics of liberal education could plausibly offer several explanations why this should be so. In "Web 2.0," Keen cites Karl Marx as a symptom of the problem and Christopher Lasch and Oswald Spengler as diagnosticians. Perhaps the contemporary university is overrun by Marxist utopians promising self-actualization to the masses. Perhaps its embrace of intellectual pluralism is a cover for an indiscriminate and licentious embrace of personal narcissism. Perhaps the crisis in higher education is an indication of broader and more baneful cultural crisis or a further indication of the decline of the West.

But there is a more likely explanation for the discord between the values of *Britannica* and those of liberal education that Keen overlooks. This explanation concerns the most problematic of all the virtues ascribed to traditional elitist culture: communitarianism. Keen quotes a 2006 speech by Jürgen Habermas to help make his case against Web 2.0: "The price we pay for the growth in egalitarianism offered by the Internet is the decentralized access to unedited stories. In this medium, contributions by intellectuals lose their power to create a focus."[65] It is worth asking who Habermas's "we" is. Society at large? Web users? the academy? intellectuals? The view he takes opposes "contributions by intellectuals" to "decentralized access to unedited stories"—informed opinion versus unfiltered mass opinion—in absolute terms. But the college classroom is precisely the place that these opinions enter into dialogue, not just occasionally but foundationally. Liberal education may grow out of the kind of print culture represented by the libraries that subscribe to *Britannica,* but it is not synonymous with print culture. Indeed it grows, adapts, and endures precisely by questioning the verities of that culture. Although their relationship is anything but symmetrical, teachers learn from their students just as students learn from their teachers. And if students learn by becoming more like their teachers, what they learn is that authority is earned and validated by scholarly communities by the ways it is performed. Students certainly earn grades and degrees by mastering areas of knowledge new to them. But they are also rewarded for learning how to think more critically, to ask better questions, and to provoke exactly the kinds of productive debate their teachers have presumably provoked within and above them.

This process requires a community in which teachers have greater power and authority than students, but one in which each student constantly aspires to greater authority by performing the roles teachers have modeled and the greater power that comes with the mastery of those roles. If the academy is a community of masters and apprentices, it is one in which apprentices are expected to advance to the mastery that will allow them to challenge, succeed, and someday dethrone the authority of their most inspiring teachers. Nor are teachers only incidentally the masters concerned with helping apprentices achieve mastery, as plumbers and cabinetmakers are; that is the most essential part of their job. Teachers understand that the bedrock concern of liberal education is not the development of the individual student but the flourishing of the communal culture teachers and students share. And if teachers succeed in their work, that understanding is one of the most vital legacies they pass on to their students.

In a presentation recorded in *The Truth According to Wikipedia,* Keen asks his audience what quality the following words share: need, want, can, should, fear. The answer that emerges is that each of them is incomplete without the preceding word "I" that marks them all as individual and narcissistic. Keen does not notice that they can all just as comfortably follow other pronouns like "they" and "you" and "we." The forces of enthusiasm, passion, ad hoc expertise, and a determination to make one's voice heard can just as readily be ascribed to communities as individuals. So can authority, as Internet analyst Clay Shirky has pointed out: "An authoritative source isn't just a source you trust; it's a source you and other members of your reference group trust together." Shirky cites Wikipedia as an instance of "algorithmic authority"—that is, "the decision to regard as authoritative an unmanaged process of extracting value from diverse, untrustworthy sources, without any human standing beside the result saying 'Trust this because you trust me.'" Algorithmic authority differs from personal or institutional authority, because "it takes in material from multiple sources, which sources themselves are not universally vetted for their trustworthiness, and it combines those sources in a way that doesn't rely on any human manager to sign off on the results before they are published." When such an authority produces good results, people come to trust it. What confirms it as an algorithmic authority is that "people become aware not just of their own trust but of the trust of others: 'I use Wikipedia all the time, and other members of my group do as well.' Once everyone in the group has this realization, checking Wikipedia is tantamount to answering the kinds of questions Wikipedia purports to answer, for that group." Hence "the criticism that Wikipedia, say, is not an 'authoritative source' is an attempt to end the debate by hiding the fact that authority is a social agreement, not a culturally independent fact. Authority is as authority does." Of course, the algorithmic authority of Wikipedia is not absolute. It is simply one more entrant on a "spectrum of authority from 'Good enough to settle a bar bet' to 'Evidence to include in a dissertation defense.'"[66] But that is equally true of all authorities, which are ultimately weighed, compared, and discounted by social groups like Wikipedia users or the academy.

A perfect example of this process of discounting is what Keen calls "the moral obligation to question the development of technology,"[67] which flourishes nowhere more vigorously than in the humanities. When Keen warns that Web 2.0 will give us less of Mozart and "more of ourselves," he does not seem to realize that that is what liberal education already promises to give students: more of themselves, not in numbers but in depth and clarity of understanding, by preparing them to take their place in a community of informed citizens who must constantly negotiate among themselves to earn the authority many of them arrived in college assuming would be their birthright. Liberal education acts best as the conservator of intellectual authority when it is most uncompromising in its criticism of authority in any form whatsoever, including that of the academy itself.

Given the surprising gaps between the principles the academy invokes to justify its strictures against Wikipedia and the principles implicit in its own practices, it seems unwise, perhaps impossible, for academics to make a case against Wikipedia based on their own institutional principles, some of which

Wikipedia shares, some of which academics do not always observe themselves. Instead, it would be both more generous and more prudent for observers and interested parties to involve themselves more fully and critically with Wikipedia, even if their ultimate goal is a more definitive rejection of it. My next chapter considers some ways to engage Wikipedia more actively and the lessons these engagements are most likely to impart.

CHAPTER FOUR

Playing the Encyclopedia Game

LEARNING MORE ABOUT HOW WIKIPEDIA works and how it compares with other reference sources can give fans and skeptics alike a firmer and subtler grasp of the problems behind its claims to authority. Teachers who want their students to examine the nature of authority itself more critically can take them further by enlisting Wikipedia as a teaching tool. Active Wikipedians—users who tinker under its hood and test its limits by examining the history of specific pages, editing pages in which they have particular expertise, and creating new entries about subjects that spark their passion—can learn much more about Wikipedia, and about their own attitudes toward authority, than mere consultant readers ever can. This chapter considers how using Wikipedia actively, productively, and repeatedly can serve the goals of liberal education even as it encourages its users to reassess those goals. Veteran Web analyst Howard Rheingold advises novice users to cultivate five new modes of online literacy: attention, crap detection, participation, collaboration, and network smarts.[1] Becoming an active Wikipedian dramatically improves the learning curve for the third and fourth of these habits and has surprising benefits for the second. This chapter, however, focuses on a sixth mode Rheingold does not mention: playfulness. Just as foreign languages are most efficiently absorbed through immersion, the best way to learn about Wikipedia is to speak Wikipedia, to do Wikipedia, to play Wikipedia.

Establishing Authority

Both critics and defenders of Wikipedia agree that the distinctiveness of its claims to authority and the apparatus it has put in place to support those claims reflect a crisis in authority posed by the recent explosion of online information, much of it unedited, unregulated, and unreliable. Wikipedia does not merely reflect this crisis but raises pointed questions that encourage deeper exploration of it. This exploration extends throughout the pages of Wikipedia and indeed the entire Web, along with bookstores, libraries, archives, schools, and universities. Caught up in this crisis, contemporary researchers can either remain increasingly fretful and uncertain consumers of information produced, published, and cited by others or investigate the possibilities of becoming authorities themselves, even though this latter course may seem presumptuous, fraught, and risky. Risk is indeed essential to the value of this task, for no one can be weaned away from an unquestioning acceptance of authority without taking risks.

Engaging Wikipedia more actively takes both teachers and students to the heart of what Richard P. Keeling and Richard H. Hersh call "higher learning," as opposed to "academic learning": "The desired learning goals of college must embrace not simply the active acquisition of knowledge, but also the active and increasingly expert use of that knowledge in critical thinking, problem solving, and coherent communication, as well as the personal, psychic, emotional, social, and civic learning of the student."[2] In Keeling and Hersh's terms, actively using Wikipedia—editing, correcting, or creating its entries—offers ideal opportunities for the kind of "apprenticeship"[3] that allows students to demonstrate both their mastery of new knowledge and their ability to turn it to use by establishing, developing, and constantly testing and redefining their own authority. This is a central aim of liberal education, though it is rarely identified as such.

Students often have difficulties mastering new knowledge actively because of their essentially passive attitude toward authority. Forget about asking us to show our work, they seem to say; just tell us what the answer is, and we'll repeat it back to you so that we can get on with our lives. The term papers traditionally assigned to high school students as exercises in gathering and synthesizing research operate in unfortunate ways to reinforce rather than complicating or transcending this deferential attitude toward authority. This is particularly likely if what the research students are asked to do is framed by guidelines that emphasize the need to avoid plagiarism by giving due credit to the authorities they consult or the number and variety of sources they must cite in order to fulfill the assignment.

An additional obstacle these term papers suggest is teachers' own attitudes toward authority, which in many cases are little more sophisticated than those of their students. Several years ago, Gardner Campbell, blogging about the 2005 Wikipedia controversy involving the vandalized passage that implicated journalist John Seigenthaler in the murders of John and Robert Kennedy, reflected:

> Now, more than ever, we need clear thinking, rigorous reasoning, about authority: its nature, purpose, and relation to justice and democracy. Teachers are of course vital researchers in this area, or should be. We conserve authority. We interrogate authority. We create authority. And we urge and encourage those capacities in our classrooms every time we convene a class.[4]

Joan Vinall-Cox wrote in response:

> Hear, hear!
>
> The question of authority is central to the websphere, but we haven't figured out how to recognize it yet. My worry is that at a basic level we are not teaching children how to evaluate the websites and information they find on the web. Many teachers and educational planners aren't familiar enough with both the web and the mindset of digital natives to help the young learn suitable critical thinking.
>
> I've read about research that evaluates Wikipedia as being more accurate than Encarta according to the experts in the 60 fields they checked—that's websay (like hearsay). Can it be trusted?[5]

Both writers are talking about the problems of teaching students about authority, but they conceive authority in fundamentally different ways. For Campbell, the nature and operation of authority are essentially problematic. Teachers have a special responsibility to conserve, interrogate, and create authority, to attend to its nature and purpose and the civic values it implies, and to help their students do these things themselves as they grapple, first in the classroom, then on their own, with problems of authority. Vinall-Cox also refers to "the question of authority," but for her the problem is simpler, perhaps because she is considering its implications "at a basic level" in which younger students are grappling for the first time with problems of authority. The mark that indicates students have mastered this level is not their more reflective interrogation of authority but their ability to rank competing authorities. Some sources (maybe Encarta, maybe Wikipedia) are more reliable than others, and students at this level, presumably the primary grades, need to learn how to recognize them. The measure of their "suitable critical thinking" is the ability to answer Vinall-Cox's final question—can the rumor about Wikipedia versus Encarta be trusted?—and all similar questions either yes or no.

The ability to determine which authorities are more and less reliable is an indispensable developmental stage in the development of students' attitudes toward authority. But it should not be mistaken for a final stage. The labors of primary teachers who struggle to make their students remember the differences between, for example, dot-com sites (those that can carry advertising), dot-net sites (those that cannot and are therefore less prone to certain obvious kinds of bias), and dot-gov sites (those created and curated by agencies of the United States government, often held up as the gold standard for the certification of authority) are superseded only a few years later by the work of secondary teachers who point out how loosely regulated all these suffixes are and how many websites are more or less reliable than their suffixes would imply. No mere taxonomy of websites can provide the certainty Vinall-Cox seeks. Nor can Wikipedia, Encarta, the whole Web, the Library of Congress, or any other authority. Indeed the whole project of liberal education might be described as weaning students away from such certainties or the quest for them, not in order to substitute an equally unnuanced and futile skepticism but rather to cultivate a more critical attitude toward authority as both a relative and an absolute value.

The crucial insight that not all claims to authority are equally valid leads to a task of paramount importance at a certain educational stage: the task of categorizing and ranking different websites, and different reference sources generally, as more or less authoritative or reliable or useful. This task, and the insights to which it leads, can carry most students through high school. By the time they reach college, however, their instructors' probing questions, and their own increasingly adventurous explorations of the Web, should encourage them to think more critically, not only about the relative authority of different sources, but about authority itself.

What's behind the Entries?

Wikipedia provides a dramatic illustration of the problematic nature of authority as well as a laboratory for sharpening students' skills in negotiating claims about authority and its fellow virtues. Learning to use Wikipedia requires the cultivation of new skills that begin with recognizing how Wikipedia works as an alternative to rather than as an unsuccessful copy of earlier reference sources and leads to developing a new kind of literacy. The first lesson in this new literacy is learning how to read Wikipedia as something different from, something more than, a series of informational pages. As journalist Cory Doctorow explains:

> Wikipedia entries are nothing but the emergent effect of all the angry thrashing going on below the surface.

> No, if you want to really navigate the truth via Wikipedia, you have to dig into those "history" and "discuss" pages hanging off of every entry. That's where the real action is, the tidily organized palimpsest of the flamewar that lurks beneath any definition of "truth."
>
> *The Britannica* tells you what dead white men agreed upon, Wikipedia tells you what live Internet users are fighting over.
>
> *The Britannica* truth is an illusion, anyway. There's more than one approach to any issue, and being able to see multiple versions of them, organized with argument and counter-argument, will do a better job of equipping you to figure out which truth suits you best.
>
> True, reading Wikipedia is a media literacy exercise. You need to acquire new skill-sets to parse out the palimpsest. That's what makes is [*sic*] genuinely novel. Reading Wikipedia like *Britannica* stinks. Reading Wikipedia like Wikipedia is mind-opening.[6]

The truth Doctorow ascribes to *Britannica,* of course, is an illusion in more senses than he indicates. The imposing look of the complete *Britannica* arrayed on its shelves, the heft and apparent solidity of each volume, and the assured tone of each entry all project a specific kind of authority, the kind intended to provide an answer for every question and brook no dissent. But anyone who has ever written for an encyclopedia knows that the content of every article, even if it ultimately presents itself as "what dead white men agreed on," emerges from a series of internal debates within writers and external negotiations and compromises between writers and editors who know all too well that "there's more than one approach to any issue." What makes Wikipedia uniquely valuable is that instead of hiding these negotiations and debates beneath an editorial apparatus and production design intended to elicit confidence in a monolithic structure whose credentials are impeccable, it makes them available to anyone with the slightest interest in consulting them.

Doctorow's argument is echoed by Danah Boyd of Microsoft Research in a 2007 talk to Pearson Publishing:

> Wikipedia certainly has its flaws, but it's not evil. In fact, it's an ideal site for learning how to interpret information. Consider California History Standard 11.1.2 where students are supposed to learn about the cultural dynamics behind the American Revolution. The view from the American and British history textbooks is quite different, yet, the English Wikipedia entry has to resolve these two perspectives. Right now, teachers say that what's in the textbook is right and what's in Wikipedia is wrong. Imagine, instead, if teachers helped students understand why these two differed. Imagine a

> culture where information is collectively valued, but youth are taught the skills for interpreting it and evaluating it rather than simply being told that everything in the information ecology that they inhabit is "bad" simply because it's not in traditionally vetted sources.[7]

Philosopher P. D. Magnus therefore concludes that "the question of whether we should trust *Wikipedia* becomes the question of *how* and *to what extent* we should trust *Wikipedia*."[8]

As Doctorow points out, the deepest truth of Wikipedia lies in the Talk and View History tabs that head every entry. Anyone who consults the notes at the end of the Wikipedia entry on George W. Bush can find the source for the verdict in the final sentence of the entry's opening section: "Although his presidency has been rated among the worst by scholars [17], his favorability ratings in public opinion surveys have improved since he left office in 2009." Note 17, which refers to an article by Kenneth T. Walsh,[9] might seem to settle the question of Bush's rating for good by providing an authoritative source. Only by consulting the Talk page of the entry can readers sample the vigorous debates among editors behind the decision to include the first half this sentence in the lead section, its balancing by the second half of the sentence, and the implication that the second half actually balances the first. In another example, the Talk page of the entry on the Ku Klux Klan presents a long argument mostly between two editors about whether it is fair to identify the Klan's beliefs, during or since Reconstruction, as "far right," given that its members at the time of its founding were more likely to be affiliated with the Democratic than the Republican Party and its principles and tactics are clearly distinct from those of contemporary Tea Party conservatives. And aspiring cultural historians who want to see rumors, denials, claims and counterclaims repeatedly reshaping a Wikipedia article can consult the View History pages for the entry on golfer Tiger Woods and read the four hundred edits in the month after the one-vehicle accident on 27 November 2009 that eventually revealed Woods's multiple infidelities, ended his marriage, canceled many of his endorsement deals, and led to his four-month withdrawal from the PGA tour.

Readers who consult these tabs can get a taste of the ways current events coalesce into history. They can follow the continuing debates behind carefully crafted language that might otherwise seem to have been cast in stone from the beginning. They can see which debates are quickly resolved and which ones continue to flourish, often to the point of tedium or farce, and why. By sampling dozens of compromises deemed necessary to resolve disagreements about even the most apparently unexceptionable articles, they can be given the equipment to reassess the commonsensical view of compromise as a formulation halfway between two opposing positions. Only by consulting the extensive Talk or View History sections of the entry on Olivia Newton-John, for example, are readers encouraged to move from the not especially interesting question of whether Newton-John—born in England, raised mostly in Australia—is British or Australian to the much more interesting question of who and what determines a given citizen's nationality, whether or not she is Olivia Newton-John.

Most important of all, readers of the Talk or View History sections can contemplate questions about the nature of authority most entries in Wikipedia and other reference sources finesse or conceal in their main entries, questions that flash out with redoubled energy in these sections. Who gets to decide whether George W. Bush was one of the worst American presidents, professional academic historians or citizens at large? Does being ranked thirty-fourth of forty-four presidents make Bush one of the worst? How much weight should be given to polls taken during his presidency, immediately afterward, and several years afterward? How should the results of these polls be reconciled with the fluctuating approval ratings routinely published throughout modern presidencies? Does having lower approval ratings than Bush at the corresponding point in his presidency indicate that Barack Obama is an even worse president? How should historians' disapproving views of the Bush presidency be framed, and how much prominence should they be given?

Questions like these are by no means unique to Wikipedia. What is unique is Wikipedia's transparency in making them readily and comprehensively available to its users. The pitched battles duly enshrined on Wikipedia's Talk and View History pages—accusations and counteraccusations of political bias, mutual recriminations concerning violations of Wikipedia's civility policy, characterizations of certain descriptions as either unfairly suppressed minority viewpoints or simple vandalism—clearly undermine Wikipedia's authority for many observers suspicious of crowdsourcing. It would be more precise, however, to say that they propose different terms for framing, understanding, and interrogating that authority. It might seem a matter of indisputable fact, for example, that Christopher Columbus arrived at what he called the New World on 12 October 1492. But although no one disputes the date of Columbus's discovery, many observers, on Wikipedia and elsewhere, dispute the appropriateness of the terms "discovery" (were the lands of the Western Hemisphere, which Norse sagas represent Leif Ericson as having colonized around 1000, and widely settled by stable populations long before then, undiscovered before 1492?) and "New World" (how new was the Western Hemisphere to anyone besides the Europeans who colonized it?). The editorial tabs in Wikipedia encourage editors and casual readers alike to make sharper distinctions between absolute truths, defensible opinions about those truths, and truths that are less absolute than they seem but still widely accepted as true.

Playing with Wikipedia

Readers who agree with Cory Doctorow in approaching Wikipedia as an arena for media literacy will follow his injunction to "parse out the palimpsest"—that is, to make sense of the layers upon layers of revisions to virtually every entry—by cultivating a less absolutist, more provisional approach to authority. Instead of accepting a single authority and defending it against all comers—or seeking a trustworthy rubric that will rank more and less reliable authorities once and for all—frequent and thoughtful users of Wikipedia are encouraged to ask the questions that will help them develop the skills to assess claims to authority themselves and ultimately to develop a new attitude toward authority, an attitude that can best be described as playful.

This may seem a strange and perverse description. Surely we owe authorities our serious allegiance, even if it is not our unquestioning allegiance. Whatever our reservations about authority, nothing can be gained by simply trivializing or flouting a given authority or the issues it raises. Bob McHenry, of the *Encylopaedia Britannica,* clearly meant his characterization of Wikipedia as "the encyclopedia game, played online,"[10] to be damning. But the alternatives of unquestioning allegiance and facile dismissal constitute a false choice, for attitudes toward authority can be playful without losing any of their seriousness. And Wikipedia everywhere demonstrates the value of seriously playful attitudes toward authority and their advantages over uncritical attitudes that may be more earnest but are also more insecure and less mature.

The specific kinds of playfulness that Wikipedia solicits and depends on are well established in studies of games and play. In *Man, Play, and Games,* Roger Caillois has defined play as free, separate, uncertain, unproductive, governed by rules, and make-believe.[11] Wikipedia clearly meets four of these criteria. Participation in Wikipedia is free in the sense of being optional or discretionary. Unlike the paid staff and consultants of *Britannica,* no one is required to create or edit Wikipedia, and contributors may opt out at any time without notice. (Even consulting Wikipedia has more the air of hobbyist browsing than the kind of work put in by researchers and students, who rarely consult print encyclopedias for fun.) Although it is constantly expanding and potentially boundless, the realm of Wikipedia is clearly demarcated from the offline world and from other online sites. The uncertainty involved in editing Wikipedia, a realm in which the work of any editor can be overwritten or reverted by other editors, is one of its most distinctive features. And Wikipedia is certainly governed by a proliferating multitude of rules.

At first blush, Wikipedia is neither unproductive nor make-believe. Its program from the beginning has been to create and disseminate a vast and authoritative base of knowledge that any user can edit. Such a project is obviously based in a world of practical exercises and results. Yet the most dedicated and industrious Wikipedia editors—like Justin Knapp, who in September 2012 became the first Wikipedian to log over a million edits—can seem to observers outside the Wikipedia bubble to have their heads in the clouds as they spend hours a day in unpaid labor. Knapp, who averaged 385 edits a day between 2005 and 2012, commented on his own activity: "Being suddenly and involuntarily unemployed will do that to you."[12] His attitude toward his work is entirely consistent with Johan Huizinga's definition of play: "a free activity standing quite consciously outside 'ordinary' life as being 'not serious,' but at the same time absorbing the player intensely and utterly."[13]

Although it might seem frivolous to characterize Knapp's work as play, a label apparently more appropriate to the vandalism with which Wikipedia is constantly afflicted, Caillois provides a helpful distinction between two different but equally playful impulses. *Ludus* is for Caillois the impulse to create and organize games according to highly structured rules, *paidia* the impulse to act playful in spontaneous, unstructured ways.[14] In these terms it might seem clear that Wikipedia vandals are motivated by *paidia*, Knapp and other editors by the distinct but equally playful impetus toward *ludus*. But the relation of *ludus* and *paidia* is more complicated than that, as E. Gabriella Coleman has shown in her study of hackers, whose activity amounts to "an expansive yet pragmatic practice of instrumental yet playful experimentation and production. In these activities the lines between play, exploration, pedagogy, and work are rarely rigidly drawn."[15]

Brian Sutton-Smith's list of rhetorical frameworks that explain why people play—play as progress, as fate, as power, as identity, as the imaginary, as the self, and as frivolous[16]—goes further to account for the nonfrivolous play, or the amalgam of playfulness and seriousness, in Wikipedia. Building the world's biggest encyclopedia may be serious business, but the activity that goes into it is not businesslike at all in a capitalist sense. The Wikipedia entry on Wikipedians cites a 2010 study that found that, "although people might initially start editing Wikipedia out of enjoyment, the most likely motivation for continuing to participate is self-concept based motivations such as 'I like to share knowledge which gives me a sense of personal achievement.' "[17]

Like many another wiki, Wikipedia was born in a spirit of play. It began in 2001 as a frankly playful adjunct to Nupedia, the online encyclopedia Jimmy Wales had launched the year earlier. Although Nupedia was free to users, it was not wiki-based but edited by experts along the lines of print encyclopedias. As a result of this peer-editing process, Nupedia posted only two full-length articles in its first six months, and Nupedia editor-in-chief Larry Sanger, who in January 2001 proposed developing new articles as wikis, hoped that the new project, whose users could freely edit its entries, would serve as a more informal laboratory and developmental site for Nupedia articles. Sanger proposed that the new enterprise be called Wikipedia, "a silly name for what was at first a very silly project."[18] Because its open architecture allowed Wikipedia to grow exponentially faster than the site it had originally been designed to feed, it ironically eclipsed Nupedia, which shut down in 2003, several months after the departure of Sanger.

Even as it has grown in scope and ambition, Wikipedia continues to be viewed as a poke in the eye to more established reference works, an exercise in virtual community building, and a utopian adventure that does its best to keep rules and enforcers to a minimum. Jimmy and Jimbo Wales, the nicknames by which its cofounder prefers to be known, are designations clearly chosen for their playfulness. The contradictions in Wikipedia's self-image as a universal reference—it wants to be both up-to-date and enduring, insists on basing its every assertion in the print sources with which it competes, and strives to be taken seriously without taking itself too seriously—stem largely from its impulses toward play. Many venues within Wikipedia explicitly foster a playful attitude, even though Wikipedia as a whole seeks to contain and channel this attitude in the service of a bold and serious project. As Joseph Michael Reagle Jr. observes, "Humor is not a policy or guideline of Wikipedia, but it suffuses the culture."[19] Whenever Wikipedia slips up and betrays its principles or reveals its inadequacy, it is typically because it has forgotten to be playful.

Wikipedia is at its most playful in discussions of its own problems and least playful when those problems are playing out in particular entries. Its View History tabs are full of disagreements and edit wars conducted in deadly earnest by all parties. The self-righteous certitude of many editors, which often descends to name-calling, mutual recriminations, and obsessive attention to the most trivial details, can lead to bitterness among individual contributors, appeals for sanctions, and condescending attention from more traditional media. Saabira Chaudhuri, for example, published a *Wall Street Journal* article examining the extended Wikipedia controversy over whether the Beatles should actually be referred to as The Beatles.[20]

The most bitterly fought of Wikipedia's edit wars are its notorious revert wars, in which dueling pairs of editors have taken turns deleting each other's revisions and restoring their own. Revert wars so damaged Wikipedia's reputation and its mission that in 2004 Wikipedia instituted the three-revert rule (3RR):

> An editor must not perform *more than three reverts* on a *single page* within a *24-hour period*. Undoing other editors—whether in whole or in part, whether involving the same or different material each time—counts as a revert. Violations of the rule normally attract blocks of at least 24 hours. Any appearance of gaming the system by reverting a fourth time just outside the 24-hour slot is likely to be treated as a 3RR violation.[21]

Certain entries judged especially volatile, like "Armenian Genocide," may be placed under an even stricter one-revert rule that blocks any editor who reverts more than one change within a twenty-four-hour period.

In practice, edit wars and revert wars have become relatively rare in Wikipedia, affecting only a small percentage of its entries. Wikipedia's adopting of 3RR and its consistent pleas that editors use Talk pages rather than edit wars to resolve their differences have sharply reduced these conflicts. But when a reference source like the English-language Wikipedia includes over four million articles, there will still be thousands of such wars, many of them waged over exactly the entries most likely to be frequently consulted and widely discussed. In "Dynamics of Conflicts in Wikipedia," Oxford researcher Taha Yasseri and four other collaborators submit a sample of intensively edited Wikipedia entries to statistical analysis and divide these entries into three categories. Some of them, like the entries on pumpkins and on Benjamin Franklin, reach quick and relatively untroubled consensus. In these entries, "usually growth starts slowly and with an increasing acceleration until it reaches a maximum speed of growth. Afterwards, when the hot period of war is passed, the growth rate decreases and consensus is reached." A second category, represented by the entry on Michael Jackson, is marked by the "sequential appearance of war and peace periods in a quasi-periodic manner. After the first cycle of war and consensus as described in (a), internal or external causes initiate another cycle. Exogenous changes happen completely randomly, but the endogenous causes may be contributed by a simple mechanism such as a constant influx of new editors, who are not satisfied with the previously settled state of the article." A third category is marked by "never-ending wars": "In the evolution of the articles in this category no permanent, or even temporary, consensus gets ever built. Articles describing intrinsically highly controversial/hot topics [e.g., anarchy, Barack Obama, Israel, and the apartheid analogy] tend to belong in this category." Yasseri et al. conclude that

> conflicts and editorial wars . . . restricted to a limited number of articles which can be efficiently located, consume considerable amounts of editorial resources. . . . Conflicts have their own temporal fingerprint which is rooted in memory effects and the correlation between edits by different editors. Finally, we demonstrated that, even in the controversial articles, often a consensus can be achieved in a reasonable time, and that those articles which do not achieve consensus are driven by an influx of newly arriving editors and external events.[22]

Given that this study finds only one percent of Wikipedia articles controversial, Wikipedia's reputation for contention and anarchy may well be overblown. More to the point, the study strongly suggests that edit wars represent not a distinctive feature of Wikipedia culture but a failure of incoming editors to adapt quickly and completely to that culture. A new editor's certainty that he[23] is right and everyone who disagrees with him must therefore be wrong is, after all, a survival from the *Britannica* mentality, which packages authority in forms that encourage unquestioning allegiance. Novice Wikipedia editors, maintaining the same absolutist attitude as they consider competing authorities who share their attitude but not their factual or ideological beliefs, are in for an awkward transitional period that ends only when they adopt a more contingent and playful attitude instead. Many editors, unable to make this transition from a work ethic of earnest rectitude to a play ethic of productive contribution, follow Wikipedia cofounder Larry Sanger by ending their association with Wikipedia and retreating in disillusionment. Their loss of faith is directed not at what Wikipedia is but at what they think it should be: a forum to display their own expertise about information, style, and formatting in order to make them feel more efficacious and self-confident.

Wikipedia is indeed a forum, but one more genuinely interactive and communal. Contributors who can step back from the precipice of individual edit wars into meta-analysis become not only more judicious and Olympian but more playful. Wikipedia's own entry on "Lamest Edit Wars," "a showcase of situations where people lose sight of the big picture" in which "humorous, insightful commentary is encouraged," lists among its hundreds of examples Frédéric Chopin ("Was Chopin Polish, French, Polish-French, or French-Polish?"), Angels & Airwaves ("more than 40 reverts in one hour by two editors. The point of contention? Whether 'Angels & Airwaves' *is* a band or 'Angels & Airwaves' *are* a band"), London Underground ("Should the term 'period' or 'full stop' be used to describe a full stop [or period]?"), and the Beatles (or is it The Beatles? and should their members be listed in alphabetical order or in the "traditional order" of John, Paul, George, and Ringo?). "Lamest Edit Wars" favors play over display in its own burlesque take on the big picture:

> *Some discussions are born lame; some achieve lameness; some have lameness thrust upon them.* Upon coming across a discussion that is borderline lame, some Wikipedians may be tempted to go do something useful. This is a big mistake. Left to its own devices, the discussion might inadvertently become useful. What's the fun in that? It is essential that as many editors as possible chime in, not adding to the discussion at hand, but merely commenting how lame it is and what a big waste of time it is. (See Self-fulfilling prophecy, Positive feedback, and Exponential growth). Merely stating the discussion is lame is frequently not sufficient; every opposing statement must be denied with increasingly vehement assertions of the lameness. While at first blush, wasting time whining about what a waste of time something is may seem illogical, the inherent irony just magnifies the lameness. An additional step to increase lameness is to include repeated links to this essay, which is WP: LAME. Administrators have a special role to play; proposing/implementing topic bans on lame participants is doubly effective: it not only increases the present-day lameness, but, by quashing debate, helps ensure the lame issue remains unsolved for future generations of Wikipedians to go on about. Lamely.[24]

As its reference to its own lameness indicates, this page, like Deacon of Pndapetzim's similarly satirical page "How to Win a Revert War"—since "on Wikipedia . . . *knowledge is egalitarian, discipline is not,*" zealous editors can prevail over both their wrongheaded opponents and "neutrals" swayed by the siren song of the neutral point of view I examined in chapter 2—may seem to provide nothing more than still another example of the whining it excoriates. "Lamest Edit Wars" begins with the admonition: "This page is intended as humor. It is not, has never been, nor will ever be, a Wikipedia *policy or guideline.*"[25] Yet all the comments recorded on the Talk page for "How to Win a Revert War" praise it as both seriously penetrating and satirically amusing.

Serious Play

Wikipedia's inclusion and framing of pages like "Lamest Edit Wars" and "How to Win a Revert War" shows how it attempts at once to encourage playfulness and to contain it within carefully demarcated safe zones—Talk pages, User pages, pages clearly labeled as humorous—so that it does not bleed onto entries that do the serious work of an encyclopedia. Commentators on Wikipedia and elsewhere have often linked its attempts to discourage or contain playfulness to "deletionists" intent on purging Wikipedia of substandard content in order to raise its standards, as opposed to "inclusionists" who urge coverage of a broader range of topics and lower standards for initial entries on the grounds that those entries can always be improved. Deletionists are motivated by conservative and self-serious ideas of what an encyclopedia should look like, inclusionists by more hypothetical, inchoate, and playful ideas of what may turn out to be important.

Despite the best efforts of editors and the sharpest observations of commentators, however, Wikipedia has most often united seriousness and playfulness in an indissoluble dialectic. The most important laboratory for this dialectic is the editing process. Even though only a small fraction of the people who consult Wikipedia make any attempt to edit it, editing is essential not only to the continued well-being of Wikipedia but to its users' relationship to it. By far the longest of the five parts of the handbook *How Wikipedia Works* concerns editing,[26] and the first and longest part of *Wikipedia: The Missing Manual,* "Editing, Maintaining, and Creating Articles," begins with the chapter "Editing for the First Time," as if editing, not reading, marked the beginning of each user's most meaningful relationship with Wikipedia.[27] Wikipedia maintains a page that encourages college teachers to design editing projects for their students, gives reasons why such projects are likely to be useful for Wikipedia and the students, sets forth guidelines for conducting editing projects with and without experienced facilitators from the Wikipedia community, and provides a brief list of contact people willing to help set up such projects and a more extensive and detailed list of current projects.[28]

In practice, however, no such advice is necessary, for anyone can edit Wikipedia without directed instruction or preparation. In a class I taught a few years ago on American mythologies, I replaced the obligatory essay on representations of General George Armstrong Custer in Hollywood films with an assignment that required students to edit Wikipedia's Custer entry or one of a number of related Wikipedia entries, by adding new information, changing what was already listed, or making deletions.

As part of its program of containing its contributors' playfulness within safe boundaries, Wikipedia invites all users who wish to propose changes to Wikipedia to post their suggestions in one of a number of Village Pumps, accessed through http://en.wikipedia.org/wiki/Wikipedia:Village_pump, soliciting comments from other members of the community before making them available to a wider audience. Along the same lines, Wikipedia's page on school and university projects advises new editors to "keep it real" by practicing their skills in the Sandbox (WP: SAND) before venturing onto the pages they plan to edit. The Sandbox does not give novices any additional help in honing their editing skills, merely a place to try them out. Because the Sandbox is frequently overwritten and erased and records no permanent changes, none of the practice sessions can have lasting consequences or provide editing lessons to anyone but the particular novice editor who is practicing.

Some of my students played in the Sandbox before moving on to the pages they planned to edit; others, following Clay Shirky's "publish-then-filter" model,[29] played on those pages themselves. Although my students were not experts on the history of the American West, none of them had any trouble editing the entries they chose. Most of their edits did not last for more than a few weeks, but they came away impressed by what they had learned about the frequency with which the pages on which they had worked were edited, the dominance of a few administrators like the self-styled "Gen. Custer" who took a playfully yet seriously proprietary attitude toward the topic, and the efficiency of those administrators in reverting constant low-level vandalism (over a few eventful hours of successive edits and reedits on 5 May 2010, Custer's middle name was given as Pokemon, he was born in 1564 rather than 1839, and he died, not at Little Big Horn but "on the moon," and in 2000 rather than 1876).

My students quickly realized that many of the reverts had been enabled not by insomniac editors but by technological devices. Bots designed to revert obvious vandalism—one of many functions these individualized computer programs have been tailored to serve—constantly prowled Wikipedia, and the watchlists that editors like Gen. Custer had created alerted them to changes to the sites in which they were most invested. Several students observed in class, "Some people really take Custer seriously," or went still further: "Do these people think they own Custer?"

Well, yes they do. Self-appointed ownership of specific articles has long been a problem for Wikipedia. The main page that discusses the subject, Wikipedia: Ownership of Articles, labels it frankly as a problem: "Believing that an article has an owner . . . is a common mistake people make on Wikipedia."[30] The mistake extends both to users and contributors whose shared attitude toward authority is literally conservative rather than playful. The terms of Caillois's game theory suggest that ownership can be usefully set against vandalism. Vandalism, born of *paidia,* is so playful that it undermines Wikipedia's mission; the sense of ownership fostered by too narrow a focus on the rules prescribed by *ludus* is so serious that it does so as well. Wikipedia has routinely sought to maintain its organizational health by developing mechanisms to contain not only playfulness but seriousness in the interests of what might be called constructive play.

Phoebe Ayers, Charles Matthews, and Ben Yates assure novice editors who contemplate starting user pages identifying themselves to the larger community that "the way of the Wikipedian is to value spontaneity on the site, not formality" but add: "When you post any serious amount of information on your user page, though, you should adopt a thoughtful approach."[31] This balance between spontaneity and thoughtfulness, complicated by the pun on "serious," faithfully reproduces Wikipedia's own ambivalent tone, which actively encourages playful participation without characterizing it as such.

Novice contributors intimidated by the prospect of editing Wikipedia are best advised to begin by expanding a stub or writing a Requested Article, for Wikipedians eager to see stubs expanded or new articles on requested topics are most likely to respond encouragingly to any attempts to do so. Even such a simple exercise as this, however, is best undertaken in a playful spirit.

Two Ways to Play the Encyclopedia Game

The importance of engaging with Wikipedia playfully is indicated by the value and limitations of a pair of classroom projects. Cullen J. Chandler and Alison S. Gregory designed the first of these projects for their students at central Pennsylvania's Lycoming College as part of their course History 232: The Rise of Islam. Chandler and Gregory's students had to complete four tasks. First, pairs of students were asked to choose topics that were either missing from Wikipedia or represented only by a stub, write a four-to-five-page research paper on these subjects using primary and scholarly sources, and then insert the results, either in a single block or in smaller units, into the appropriate entries. Next, individual students were asked to make one small change to any Wikipedia article of their choice. Then, students were asked to document and add a single reference to a Wikipedia article of their choice. Finally, students were required to monitor the entries they had created or modified to see how later editors treated them.

Chandler and Gregory's carefully designed exercises led their students through a series of illuminating, if sometimes humbling, adventures and discoveries. The username one pair of students had chosen, "thejesuschrist-vampirehunters," was immediately censored by Wikipedia. Another pair was banned for seventy-two hours after they repeatedly posted copyrighted song lyrics on the "History of Istanbul" page. A third pair found their article on Sharia attacked by profane vandalism. But the authors of a fourth paper were gratified to see a fact from their article "Islamic Civilization during the European Renaissance" featured in the "Did you know . . ." box on Wikipedia's main page six days after they created their own page.

After the unit was complete, both teachers assessed the exercises they had designed as largely successful in helping their students become better researchers, making them more aware of Wikipedia as a research tool, giving them greater acquaintance with wiki-tools, and becoming more informed "information consumers." "By the end of the project," they reported, "the majority of students in the class (roughly 80%) said that they now thought Wikipedia was less useful than they originally thought, but that it is still a good place to find citations directing readers to usable sources."[32] Phrases like "usable sources" and "information consumers" make it clear that Chandler and Gregory wanted their students to participate more actively in Wikipedia but were not disappointed that the experience diminished their respect for it. The lesson that their students ended up learning was that Wikipedia has its place but that its authority is far less than that of traditional scholarly sources.

The student responses Chandler and Gregory record from a climactic class session in which students shared their reactions to the online treatment of their creations and additions go far toward explaining which questions these exercises did and did not raise. The most widespread reaction was

> indignation—how dare someone make changes to our article?! One student group referred to the article's other editors as "Wikijerks." For the few articles that saw no changes by others, those students now consider themselves to be the world's foremost experts (at least through Wikipedia) on the subject. This brought a fair amount of amusement to the class at the expense of said "experts," making the students contemplate once again the value of authority. Would you cite research done by your classmate?[33]

In other words, the students swiftly assumed ownership of what they considered their articles, made repeated attempts to revert later edits (in one case getting banished from Wikipedia for three days because they kept reposting copyrighted material), and dismissed competing editors as Wikijerks.

Despite the claim that the exercise led the students to contemplate the nature of authority, their conclusions seem to be exactly the same ones they had before the exercise: in every contest for authority, there are winners and losers because some people have a right to claim authority and others (like "your classmate") do not. Cullen and Gregory never encouraged their students to reconsider their assumptions about authority in light of experiences that had highlighted the often playful negotiations behind authority, presumably because they shared these assumptions themselves. So their experiment ended by confirming students' beliefs that authority was absolute in its claims, that authorities were either categorically trustworthy or not, and that the most critical attitude students could reasonably adopt toward competing authorities was to rank them in an unchanging order. In light of the deeply conservative student assessment of Wikipedia with which Cullen and Gregory end their essay—"It's okay for the layperson to get an overview, but it's not good for research unless you just use it for the references"—the single most interesting student comment they quote is the reaction of initially fearful students to the assignment once they had learned the rudimentary mechanics of editing Wikipedia: "this isn't such a big deal."[34]

Robert E. Cummings has described a very different project he conducted with his writing students at the University of Georgia. The unit he describes, which focuses on the requirement that each student "contribute to film pages in Wikipedia,"[35] included six writing assignments over a period of two weeks. Cummings's students began by reviewing Wikipedia's general goals and policies and the rhetorical strategies of the most successful film pages. Working either individually or in small groups, students developed proposals for making substantive changes to a specific film page and pitched them to the whole class, then entered those changes onto that page and monitored the results.

Although it shares certain elements with Chandler and Gregory's classroom exercises, Cummings's experiment with Wikipedia differs in important ways. He directed his students toward film pages, where the stakes would be lower and the tone more playful than in the page on "Islamic Civilization during the European Renaissance." In choosing a topic that was not directly related to the nature of the course, he increased the odds of tapping into his students' personal passion and expertise, making it clear that the focus of this assignment was Wikipedia, not the subject matter of his class. Cummings required his students early on to define exactly what they thought made some Wikipedia pages better than others. In order to head off any claims to ownership on the pages to which they posted additions and changes, he asked them to consider Wikipedia a platform for writing whose protocols they were obliged to respect. Unlike Chandler and

Gregory, he carefully prepared his students, both by requiring them to review Wikipedia's protocols in detail and by warning them himself, for the "high probability that what you write will be changed by another person on Wikipedia. Don't get upset."[36]

More important, Cummings's ultimate goal was not to attack Wikipedia's authority, or even to reconsider it more critically, but to give his students experience in CBPP (commons-based peer production), a concept he borrows from legal and economic theorist Yochai Benkler.[37] Indeed, Cummings reflects, "this study had been structured specifically to avoid these issues; within class, not much discussion was given to whether or not Wikipedia was successful or academically sound."[38] Instead of focusing on assessing the authority of specific Wikipedia entries or Wikipedia generally, students were invited to establish their own authority within the context of Wikipedia protocols they had to review and assimilate before they sat down to write.

Cummings provides detailed comments on three revision projects his students undertook to the pages of *Good Will Hunting, The Color Purple,* and *Star Wars Episode V: The Empire Strikes Back.* One contributor to the first of these projects emphasizes her impatience with what Cummings calls Wikipedia's "perceived lack of authority": "Writing on Wikipedia . . . did not really improve our writing because we were not critiqued on it. Even if someone edited your passages, you still would not know if it was a middle-schooler surfing the Internet and changing it for fun, or whether it was someone who was skilled in the subject we were writing on." Cummings observes that this contributor, valuing *logos* over *lexis,* the word of an identifiable speaker over the word of an anonymous writer, "requires a direct value judgment from an instructor on her text before she can see value in the exercise."[39]

A rift arose between the two students who created a Wikipedia page on *The Color Purple* when one of them deleted a Quotes section that the other had included. Cummings notes that this action "marks the fullest assimilation of a writer into the network," as opposed to the typical reaction students had when their peers' text was removed: "Students emotionally supported each other, often fostering an 'us versus them' mentality from within the class." He concludes that the second student's action, although it indicates a rejection of the first student's work, "marks a level of collaboration within the network that is often difficult to reach in the composition classroom itself," where "traditional peer review, when conducted in a face-to-face environment, offers almost no incentive for students to risk offending others by deleting their text."[40]

One of Cummings's most successful students was the one who contributed to the page on *Star Wars Episode V: The Empire Strikes Back* by augmenting the Radio Drama, Theme, and Setting sections and creating Awards and Nominations, Music, and Quotes sections. Initially intimidated by the community that had already created the page, she ended her comments on the entry's Discussion page by integrating her additions into what Cummings calls "the overall mission of the project in hopes of protecting her work from removal" by saying: "I added these things to make the Wikipedia entry more complete and informative. I hope my contributions helped." Although many of her additions were deleted, a veteran user responded enthusiastically to this last wish: "Yes! They certainly did. Now we can apply that to the other five *Star Wars* film articles."[41] This experience confirmed the student's preference for *lexis* over *logos*. As Cummings puts it:

> The increased audience feedback in the transactional rhetoric environment is not merely desirable in terms of pedagogy, but essential. . . . A student may well feel a sense of accomplishment after receiving praise from a classmate during a peer review, but positive feedback from a knowledge community on a CBPP contribution develops a sensibility within the writer of having produced work valuable to a larger community and immediately recognized as such, creating the desire for similar accomplishments within and beyond the education environment.[42]

Here and elsewhere, Cummings makes it clear that his unit was designed not as a test of Wikipedia but as a test of CBPP. Because "students who most seriously questioned the ideas behind open, collaborative, online writing were less successful in contributing to it," one of his primary conclusions is that "CBPP needs to be justified to students as a legitimate part of the composition writing experience." Another conclusion is that "audience awareness figures prominently in the minds of students when they compose in CBPP"; most students "indicated that they felt pressure from having to write in such a public arena." In addition, Cummings regretfully concludes that "contributors in the CBPP network have 'bought in' to the premise of their own work" by being allowed to select their own projects and choose how much time and energy they will devote to their chosen projects; his unit "erred . . . in not allowing them to 'buy in' similarly to the concept of Wikipedia itself."[43]

Playful Collaboration

Unlike Chandler and Gregory, Cummings invited his students to focus on Wikipedia as a platform for refining and extending their own authority, not as an authority to be accepted, questioned, or dismissed itself. His students' reactions to the mechanics of editing Wikipedia could usefully reflect on the project of establishing one's authority as well, not because conflicting claims to authority are trivial, but because they are incessant, subject to constant renegotiation, and independent of students' initial assumption that there is an unbridgeable gap between their efforts and those of the duly constituted authorities. Authority is at play in Wikipedia, as it is everywhere else, precisely to the extent that it is in play.

When I asked my own students to edit the pages concerning General Custer, the battle of Little Big Horn, and their cinematic representations, I framed the assignment not as an exercise designed to produce winners and losers but as an experiment designed to expose their ideas to a wider audience and see what happened next. The result was not to constitute any of them experts in Custer but to demystify the authority of Wikipedia and its contributors, of whom 27 percent are under twenty-one, and 13 percent are high school students.[44] Instead of learning that they were or were not authorities on General Custer themselves, my students learned that authority was hypothetical and stipulative, something that could be earned only by being repeatedly risked, challenged, won, reframed, lost, and won again. I encouraged them to treat editing Wikipedia as a game with real-world consequences that might seem daunting but were hardly overwhelming.

This game, I found in further experiments, was a great deal more rewarding when it was played collaboratively. After noticing that the Wikipedia entry on Ed McBain, the mystery author who created the 87th Precinct police procedurals, listed only one of his three wives, I challenged freshmen in a class that was reading *Cop Hater,* McBain's first novel, to identify his two other wives so that we could add them to the entry. The students, working individually, spent the better part of a class period seeking this information online but came up empty. The next time I asked the same question, my students were adult grade school teachers, fellows in a seminar on online research who trusted each other and were used to working collaboratively with their own students and each other. Within a few minutes, armed in part with resources that had not been available to my earlier students, they began to call out developments, then worked spontaneously together to refine their discoveries into results, analyze the authority of the online resources they had consulted, and brainstorm the best ways to enter the results in the Wikipedia entry. Although only one of the fellows could record the edits, under a username and password she created on the spot and shared with the rest of the seminar, the edits were truly a group effort, and one that different members of the seminar delighted in checking periodically over its remaining weeks. Working as editors, these fellows found, was a game that empowered them to become authorities themselves, testing their own authority by putting it in play, engaging in dialogues with other authorities that might well go disastrously wrong, sometimes crushing their self-confidence or giving them an unearned confidence, but ultimately leading them to respect and discount other sources of authority as they discounted and respected their own.

Wikipedia's signature innovation might be described as the attempt to diffuse the mutual trust that seemed naturally to arise among contributors sitting in the same room who were well-known to each other over the vast cybercommunity of Wikipedia users and editors, most of whom will never meet each other. As Joseph Michael Reagle Jr. has observed, "Wikipedia is both a community and an encyclopedia," and faith in the community "of which it is only the most visible artifact" not only makes Wikipedia a more powerful and authoritative resource but also encourages a more playful attitude toward an activity that is essentially collaborative rather than competitive.[45] Collaborators accept the authority of other contributors to edit the findings they publish as the price for their own admission to the circle of authorities entitled to put their ideas in play. Collaboration offers more low-stakes opportunities for feedback, more frequent opportunities for course corrections, and more rewarding opportunities for validation. No wonder it feels more playful than solitary labor, and no wonder Wikipedia does everything it can to foster a sense of collaboration, from urging warring editors to move their discussions to Talk pages to providing contact information for consultants willing to help classes with editing projects.

It does not follow, of course, that all collaboration is play. War is not play. The literally earth-shaking consequences of its labors made the Manhattan Project anything but playful. But the war games armies frequently stage within their own ranks are playful, as their label attests. Their participants pretending to be on opposite sides may not be grinning and slapping each other's backs throughout the exercise, but there is every chance they will do so once it is over. Even the maneuvers a nation's armed forces stage in preparation for hypothetical conflicts are more playful than those conflicts themselves for the same reason that all rehearsals are likely to be more playful than the performances for which they prepare. Every contribution to Wikipedia is a rehearsal. Even the most extended and closely reasoned additions or edits are hypothetical, experimental, and subject to change. Indeed accepting the likelihood of further change is essential to the culture of Wikipedia, unlike the culture of the Manhattan Project, whose results could never be reverted.

Playful Consultation

It is not simply collaborating on Wikipedia that is playful. Consulting it is playful as well in the sense that it is likely to be an activity more free, separate, uncertain, unproductive, governed by rules, and make-believe than consulting print encyclopedias. Partly because the scholarly community has taken such a strong stand against it, Wikipedia is a low-stakes authority often consulted on the fly to satisfy users' curiosity, check facts, refresh memories, settle bar bets, explore tendentious positions, and browse hotlinks. Even though savvy users do not trust Wikipedia to be absolutely authoritative, they do trust it. Their trust is aptly characterized by Ronald Reagan's mantra about arms negotiations with the Soviet Union—a dictum translated from the Russian and borrowed, if Wikipedia can be believed, from Vladimir Lenin—"Trust, but verify." This dictum may seem blankly self-contradictory. After all, if you trust something, why would you need to verify it? But the contingent, limited kind of trust it prescribes is both more useful and more widespread inside and outside Wikipedia than the unquestioning trust in absolute authorities of the sort with which Wikipedia is often rhetorically contrasted.

Using Wikipedia in any capacity—consulting it, contributing to it, editing it, teaching with it—encourages trust-and-verify attitudes that are more likely to produce both a more critical tropism toward conscientious verification and a more mature brand of trust. Experienced users of Wikipedia can develop the ability to weigh competing claims by discounting them according to the authority of their sources, none of which could ever be absolute. The culminating wisdom of Kenneth Burke's rhetorical analysis of not only individual utterances and motives but whole systems of thought is to "enable people to *be observers of themselves, while acting*"—to discount every authority, including one's own, by considering its source and accepting the certainty that it will be parsed, modified, and discounted by later authorities. Hence Burke concludes that "whatever poetry may be, criticism had best be comic."[46] This conclusion is the wisdom of liberal education as such. For what is liberal education but a seriously playful series of encounters with earlier authorities in which we aim to question, discount, and selectively absorb them in order to develop a more critical and confident sense of our own authority and the authority of the groups within which we think and speak and act?

When *Encyclopaedia Britannica* sought to market an adjunct product, a shelf-load of fifty-four closely printed volumes immodestly labeled the Great Books, in 1952, the set began with a long introductory essay titled

"The Great Conversation" whose author, former University of Chicago president Robert M. Hutchins, observed:

> The goal toward which Western society moves is the Civilization of the Dialogue. The spirit of Western civilization is the spirit of inquiry. Its dominant element is the *Logos*. Nothing is to remain undiscussed. Everybody is to speak his mind. No proposition is to be left unexamined. The exchange of ideas is held to be the path to the realization of the potentialities of the race.[47]

This passage sounds anything but playful. Hutchins was a deeply conservative thinker, an ideal person to introduce a project like the Great Books, whose claim to authority was equally conservative. Shorn of its uncritical celebration of Western civilization and the potentialities of the race, however, the ongoing program he memorializes and announces for putting ideas in play could readily serve as a program for Wikipedia as well. The goal in both cases is not merely the preservation but the development of human knowledge. The means are the active engagement with earlier authorities in order to develop our own. The gravitas of both programs cannot be achieved without testing, risking, creating, and recreating, in both senses of the word. Wikipedia reminds us what it suited *Britannica* to forget: that the great conversation and all the lesser conversations from which it draws are fundamentally playful.

CHAPTER FIVE

Tomorrow and Tomorrow and Tomorrow

EVERYONE AGREES THAT WIKIPEDIA, now entering its teen years, cannot last indefinitely in its present form. It is very much a work-in-progress whose future is too shrouded in mystery to say for certain whether it will die as definitively as a tragic hero or morph into another form we have not imagined. But, even if we cannot predict one clear path forward, the different possible lines of future development illuminate several aspects of its relationship to the critical thinking liberal education is designed to foster. This closing chapter is not meant as an exercise in prophecy but as a series of frank and sometimes mutually exclusive speculations that sketch out several possible tomorrows for Wikipedia and trace their implications both for the future of liberal education and for authority as such.

The Decline and Fall of Wikipedia

The easiest tomorrows to foresee are two contrary futures that are both regularly predicted: complete failure or overwhelming success. Wikipedia could fail for a wide range of reasons. The Web could be commercialized in ways that would make its bandwidth prohibitively expensive. The Wikimedia Foundation could pass into the custody of dictators less benevolent than Jimmy Wales. The site could be overrun by vandals or malign bots who turn its pages to gibberish or sabotage enough of them in more subtle ways to undermine its authority.

One powerful basis for such doomsday scenarios is ecologist Garrett Hardin's influential theory of "the tragedy of the commons." Whenever privately owned livestock are allowed to graze on public land, Hardin observes, the land is eventually overgrazed, because the immediate positive utility to each owner of adding more animals outweighs the negative utility of the land's eventual depletion. Hence "ruin is the destination toward which all men rush, each pursuing his own best interest in a society that believes in the freedom of the commons. Freedom in a commons brings ruin to all."[1] The very freedom Wikipedia accords its users, according to prophets who echo Hardin, will inevitably ensure its failure.

Against Hardin's commons, destined to destruction at the hands of selfish individuals, stands the analysis of Yochai Benkler, who projects a very different fate for the commons rooted in the emergence of a "networked information economy" that has supplanted "the industrial information economy" dominant throughout the past 150 years. Benkler singles out three factors as crucial to this new economy: the increased importance of "non-proprietary strategies" for information production; the rise of "non-market production" made possible by dramatically increased online connectivity; and "the rise of effective, large-scale, cooperative efforts—peer production of information, knowledge, and culture."[2] For Benkler, "the networked environment makes possible a new modality of organizing production: radically decentralized, collaborative, and nonproprietary; based on sharing resources and outputs among widely distributed, loosely connected individuals who cooperate with each other without relying on either market signals or managerial commands. This is what I call 'commons-based peer production.' "[3]

The future Benkler foresees for the commons is based on strikingly different assumptions about human nature than Hardin's economic model makes. Benkler argues that altruism is not always trumped by selfishness, especially when it can be encouraged at little personal cost. The most widely noted product of commons-based peer production, or CBPP, is open-source software. But as Benkler adds, and as Robert E. Cummings argued in the classroom experiments with Wikipedia I summarized in chapter 4, another such product is Wikipedia, "one of the most successful collaborative enterprises that has developed in the first five years of the twenty-first century."[4]

Wikipedia Rampant

Instead of being doomed by selfishness, perhaps the commons will release new energies in the human community that will transcend selfishness in a way that will make Wikipedia rampant, extending its reach so dramatically that it becomes a one-stop source for information of all kinds, a virtual monopoly. Given the vast amount of material archived online, of course, Wikipedia is unlikely to become a true monopoly without any rivals. But the extent to which search engines like Google and Bing already list entries from Wikipedia at or near the top of their lists of search results already means that many casual online researchers stop searching after perusing these entries, giving Wikipedia a de facto monopoly whose power increases from year to year.

Wikipedia has regularly sought to extend its empire by allied ventures. The Wikipedia home page lists eleven such projects in addition to Wikipedia: Wikibooks, Wikiversity, Commons, Wiktionary, Wikiquote, Wikivoyage, Wikidata, Wikinews, Wikisource, Wikispecies, and MediaWiki. Several of these were clearly designed to compete with other ventures: Wikibooks with Google Books, Wiktionary with Dictionary.com, Wikiquote with Quotationspage.com, and Wikinews with any number of other online news outlets. The foundation's latest creation, Wikivoyage, is an obvious competitor for Tripadvisor.com. After listing five long-range "strategic priorities"—"Stabilize infrastructure, Increase participation, Improve quality, Increase reach, Encourage innovation"—the foundation's Strategic Plan outlines five specific targets for 2015:

- Increase the total number of people served to 1 billion
- Increase the number of Wikipedia articles we offer to 50 million
- Ensure information is high quality by increasing the percentage of material reviewed to be of high or very high quality by 25 percent
- Encourage readers to become contributors by increasing the number of total editors who make at least five edits per month to 200,000
- Support healthy diversity in the editing community by doubling the percentage of female editors to 25 percent and increase the percentage of Global South editors to 37 percent[5]

The tendency of all these goals is clearly indicated by the prominence of the word "increase." The foundation wants to increase the number of entries, the number of users, the number of editors, the percentage of material reviewed and pronounced high quality, the proportion of editors who are female and resident in the Southern Hemisphere. The future the foundation envisions for itself is just like its present, only more so.

This future may sound preposterously utopian or megalomaniacal. But it is eminently consistent with the strategic plans of earlier generations of cyberspace pioneers. In the days before the Web, IBM thought its introduction of the personal computer in the early 1980s would drive out upstarts like Radio Shack and Compaq. The corporation was half-right. The upstarts wilted, but IBM's hardware monopoly never materialized. Instead, the standardization of its DOS operating system enabled the system's designer,

Microsoft, to gain a foothold it used to establish a software monopoly. In the early days of the Web, Microsoft succeeded in making its browser, Internet Explorer, dominant over the pioneer browser Netscape Navigator by integrating it with its Windows operating system, ensuring that it would be installed on most computers around the world. More recently, Comcast, through its purchase of Time Warner Inc., has sought to establish an integrated monopoly that would combine the production of news broadcasts and feature films with its online global distribution network. The success of Apple's iPod and iPhone used the public's appetite for downloading music to make Apple, a minority player in desktop computers, into a major power in the drive to shrink computers to pocket size. Google attempted to use the overwhelming dominance of its search engine as a basis for a similar dominance in the archiving of online text and is now creating devices to read that text and more. The approaches to the media may concentrate in different places—hardware, operating software, browser software, search engines—but the goal remains remarkably similar. The immediate future of cyberspace seems likely to be determined by a series of struggles among hardware, software, content, or management providers for monopoly control.

So there is nothing untoward about Wikipedia's analogous drive to extend its own reach. Many of these earlier quests for cybermonopolies, of course, failed to achieve their goals and sometimes put the questers out of business. Wikipedia's bid may end in failure as well. As Joseph Michael Reagle Jr. points out, however, Wikipedia has so far been remarkably resilient in the face of prophecies of its doom, which have changed their shape but not their frequency ever since it was launched. "Just when arguments that Wikipedia would never amount to anything ceased," observes Reagle pointedly, "new arguments about its death took their place."[6]

What makes predictions about the project Wikipedia is pursuing especially difficult is a pair of paradoxes. One is its dependence on wikis that anyone can edit, augment, or delete—a diffusion of power that seems the very opposite of the way monopolies usually work. Hence the Wikimedia Foundation can plausibly describe its vision of Wikipedia rampant as a triumph not for a small nonprofit organization with a remarkably low number of paid employees but for "the people" who make and use it. A second paradox points in the opposite direction: toward Wikipedia's sources. Since a founding principle of Wikipedia is its proscription of opinion or original research, every statement it makes, apart from those universally accepted as true, must be sourced, or capable of being sourced, by reference to print media. Wikipedia is a parasitic metapedia whose continued vigor, currency, and reliability depend on those of its sources. If Wikipedia establishes a monopoly on anything like its current terms, it will be a monopoly that depends on both the rival sources to which it defers and the universe of users and editors who maintain it.

Not Quite Wikipedia

Three factors could well alter this future. One is the open access movement in scholarly publishing, which works to archive peer-reviewed research online in addition to, or instead of, printed publications. Wikipedia already allows citations and links to other web pages in its articles and bibliographies but notes that "when available, academic and peer-reviewed publications, scholarly monographs, and textbooks are usually the most reliable sources."[7] A wholesale shift to open-access publishing would presumably soften this implied prejudice in favor of print sources. At the same time, a dramatic rise in the online access to original, peer-reviewed scholarly work might well put pressure on Wikipedia to reconsider its NOR policy, since it would be surrounded by online sources making the results of cutting-edge research readily available.

In addition, it is possible that Wikipedia will change its decision to forgo advertising, a development that has often been predicted, though less frequently in recent years. Users who noticed that Wikipedia was carrying advertisements for Chrysler or Budweiser might well take its entries on automobiles or beers with a grain of salt. But there is no reason to assume that accepting advertising or becoming a for-profit venture would compromise Wikipedia's pursuit of its stated goals any more than *Britannica*'s marketing by door-to-door sales representatives at premium prices was held to compromise its authority or public broadcasting's "underwriters" detract from its mission and outreach.

More generally, Wikipedia's volunteer culture could be overtaken by a system in which articles were routinely written and edited by paid contributors whose agendas differed subtly or sharply from Wikipedia's stated goals. So many such contributors are already active on the site that as this volume was going to press, the Wikimedia Foundation Legal Department had proposed an amendment concerning Wikipedia's Terms of Use, which "already prohibit engaging in deceptive activities, including misrepresentation of affiliation, impersonation, and fraud."[8] If, after a period of comment by the Wikipedia community, the Wikimedia Foundation Board of Trustees accepts the amendment, any contributors who receive compensation for their services from employers outside the Wikimedia Foundation will be required to disclose this information on their user pages, on the Talk pages of any articles they have been paid to create or revise, or in the edit summaries accompanying any such articles. Whatever becomes of this particular amendment, the conflict between Wikipedia's community culture and financial interests bent on shaping the information available on Wikipedia seems likely to continue.

An alternative future for Wikipedia depends on sharply revising its stated goals. A policy of "flagged revisions" first instituted in the German-language Wikipedia in 2009 and accepted for use in English-language Wikipedia entries beginning on 1 December 2012 allows authorized users to "tag" a given entry in a given form that is then made available to later users in preference to more recently edited versions until they in turn have been approved. The Wikipedia entry on Flagged Revisions lists several different ways members of the community have proposed to use this new capability. Some community members have suggested marking a specific version of a given entry as "sighted" and free of vandalism. Others have proposed tagging a specific version as having gone through "a quality assurance process" determined by a consensus of the editors. Others have advocated delaying the appearance of all new revisions for two hours, in order to give editors a chance to revert vandalism before it appears online.[9] Still others have recommended establishing "reliable" protocols enforced by empowering two separate groups: "patrollers" and "reviewers." "Patrollers," according to this recommendation, can combat vandalism and maintain quality by rating specific aspects of a given entry as accurate, moderate, or good but cannot downgrade entries above those levels. The more powerful "reviewers" can strive to improve the quality of entries by rating those same aspects up to the level of featured and can downgrade aspects of article revisions to any level.[10] In every case, however, the underlying principle, which has long been debated, involves substituting a hierarchy of edits maintained by a hierarchy of reviewing and editing authorities for the radical equality among edits and editors that has been a hallmark of Wikipedia from the beginning. Guided by this vision, Wikipedia would become more like *Britannica* by becoming less like a series of wikis and more like a series of duly reviewed, approved, and archived texts.

Still another possibility is that Wikipedia could break up or fork into multiple competing projects. Mathieu O'Neil has sketched this future in cautionary terms: "Wikipedia's lack of a constitution, or of clearly defined voting procedures that would enable this constitution to be updated, means there is a danger of the project fragmenting into a multitude of smaller wikiprojects—local jurisdictions over which a limited number of participants will have a say, and who may start writing rules that conflict with others."[11] But history offers no evidence that citizens who acknowledge only a single moral or epistemic authority are significantly better off than citizens who juggle allegiances to competing authorities. What will become of Wikipedia if it ends up as one among many online encyclopedias?

Wikipedia among the Pedias

The history of the Web is rich in curated collections of such texts whose proliferation suggests a future for Wikipedia as one of an indefinite number of variously authoritative competing reference sources. Encarta, the online encyclopedia first marketed in 1993 as a reference that added online media resources to the strengths of traditional encyclopedias, remained updated and available, though at steadily diminishing prices, until 2009. Any user can propose edits to articles in Scholarpedia.org, a peer-reviewed collection of articles focusing on physics, neuroscience, and computational intelligence, but all such edits must be approved by the expert curator of the article in question. Wikipedia itself traces its roots to Nupedia, the expert-curated site Jimmy Wales and Larry Sanger launched in 2000, for which Wikipedia was originally viewed as "a breeding ground for content to be eventually moved into the commercial Nupedia"[12] and which ended up as a source of many entries absorbed into Wikipedia instead. Citizendium.org, the venture over which Sanger has presided since 2006, is a much smaller rival encyclopedia that anyone can edit but only under his or her verified real name.[13] Scholarpedia and Citizendium's model of allowing the general public to edit articles that remain under the jurisdiction of expert curators has been adopted as well by other specialized resources like Medpedia.com, which allows any user to ask questions and suggest changes to its articles but permits only users who have been verified as holding an MD or PhD to edit the articles.

Two other models have sought to combine the open-access structure of wikis with central oversight. One is represented by Enciclopedia Libre Universal de Español, which was founded in 2002 as a fork of Spanish Wikipedia by contributors unhappy with the older versions of MediaWiki available to non-English Wikipedias and fearful that Wikipedia, a dot-com rather than a dot-org site, would begin to solicit advertising. The other is indicated by Veropedia.com, a for-profit venture which served from 2007 to 2009 as a mirror site that archived stable versions of Wikipedia articles that could not be further edited. This list might be rounded out by a selection of the print encyclopedias that are available in online versions.

Sources like these, some of them predecessors or spinoffs of Wikipedia, might be considered complements or alternatives to Wikipedia. Other sources, better described as niche or specialty encyclopedias, make stronger claims to authority within a specialized discipline. Like the venerable print *Encyclopedia of Philosophy* or *Grove's Dictionary of Music and Musicians,* the Animal Diversity Web[14] is maintained and curated by traditionally credentialed experts, although students can, given permission, both add to its entries and use its Quaardvark software to search its lists and test their hypotheses about patterns among different animals and animal behavior. Entries in the Internet Encyclopedia of Philosophy[15] are peer-reviewed and curated by experts. Like Wikipedia entries, they are designed to summarize rather than advance knowledge in their field, but they are more stable than Wikipedia entries. The Encylopaedia Biblica[16] and the Jewish Encyclopedia[17] archive century-old editions of notable printed encyclopedias, but the Catholic Encyclopedia[18] is maintained online and also marketed on CD-ROM. The Encyclopedia of Mormonism[19] is available online in both pdf and wiki format, a feat made possible by the fact that, although the entries have the same tabs as the analogous entries in Wikipedia, most of them are rarely edited or discussed. The list of specialized encyclopedias could be extended indefinitely, from the Encyclopedia of World Problems and Human Potential, available in hard copy, CD-ROM, and online,[20] to the Rocklopedia Fakebandica, a reference guide to fictional rock bands whose opening page entreats aspiring editors: "Keep it real.—Please only bands/musicians from published sources. There's plenty out there; you don't need to make up any on your own."[21]

Still other sources like Conservapedia and Metapedia, "an electronic encyclopedia which focuses on culture, art, science, philosophy and politics"[22] that "gives us the opportunity to present a more balanced and fair image of the pro-European struggle,"[23] are best described as ideological correctives to Wikipedia. Still others, like en.Uncyclopedia.co, Encyclopedia Dramatica.se, and Wikiality.wikia.com, are parodies of Wikipedia, the first a whimsically close imitation in its formatting, the second considerably more free-wheeling and foul-mouthed, the third an outlet for Stephen Colbert's distinctive brand of satirical truthiness. Soon after Nicholson Baker, reviewing John Broughton's *Wikipedia: The Missing Manual,* called for a "Wikimorgue" of articles deleted from Wikipedia—"We could call it the Deletopedia"[24]—Deletionpedia[25] began archiving such pages. The new archive quickly reached 60,000 entries that stretched from the George Nethercutt Foundation, deleted after one day because its creators had failed to establish the foundation's importance, to List of Virgins, deleted in 2006 after a dispute over the appropriateness of including Britney Spears led to a consensus that a NPOV list would be impossible to maintain.

Andrew Brown has described Wikipedia as "a cancellation of the old encyclopaedic ideal. A traditional encylopaedia, paradoxically, prided itself on its inclusivity, but was *essentially* exclusive."[26] Paul Graham implicitly agrees with this description but dissents from its air of judgment in his list of thirty "Startup Ideas We'd Like to Fund," which characterizes the deletionists Baker has battled as "constrained by print-era thinking" and concludes, "There is room to do to Wikipedia what Wikipedia did to Britannica."[27] James Gleick, reviewing the conflict after briefly describing the work of the Article Rescue Squadron, whose participants try to improve poorly written Wikipedia articles in danger of deletion, concludes: "Unlike trivial edit wars, the battle between deletionists and inclusionists has no clear path to peace, because ultimately both sides have a claim on Wikipedia's core values. The deletionists carry the banner of quality, of verifiability, of trust. The inclusionists say that more information is always better: Any article that can be deleted can be improved instead."[28]

Not only Metapedia but Conservapedia, Uncyclopedia, Encyclopedia Dramatica, Wikiality, and Deletionpedia are all metapedias, encyclopedias whose foundational impulse is their reaction to other encyclopedias. Metapedias seek to correct the limitations of other sources by revealing their political or economic agendas, archiving their discarded material, or lightening their self-serious tone. The politically crusading tendencies of metapedias are best represented by SourceWatch.org (formerly Disinfopedia), in which the Center for Media and Democracy, which also publishes PRWatch.org, BanksterUSA.org, and ALECexposed.org, monitors political discourse in order to unmask disinformation by identifying its sources and discounting them.

Two metapedias devoted to Wikipedia illustrate the range of forms metapedias can take. Wiki-Watch, a project of the Study and Research Centre on Media Law at Frankfurt's European University Viadrina, allows visitors to see the number of contributors, edits, references, and outside links in a given Wikipedia entry and the number of visitors it has attracted over the past hour, day, week, or month. On the basis of these statistics and other information described in its blog,[29] Wiki-Watch rates each Wikipedia entry from one to five stars. WikiTrust, a free software add-on to the Firefox browser,[30] adds a new "WikiTrust" tab to each version of each Wikipedia page that displays less frequently edited passages from the page against an orange background. The less frequently edited (and therefore, presumably, the less trustworthy) the passage, the brighter the orange. In addition, WikiTrust provides users with tools designed to identify which edits are most likely to be vandalism and to distinguish edits by contributors known to be reliable from edits by novice or inexperienced contributors. WikiTrust acknowledges that

> the algorithms implemented in WikiTrust cannot discover "truth," and cannot discover false information when all editors and visitors agree with it. If an article contains an error, and the error goes unnoticed in multiple revisions, WikiTrust will be of no help in identifying it. This is similar to the process of scientific review: once the reviewers agree, papers are published in conferences or journals, and the occasional error can (and often does) slip through.[31]

Although WikiTrust's status as a metapedia may not be as clear as that of Wiki-Watch, they both share the goal of all metapedias. In the attempt to distinguish more and less authoritative statements in these sources, metapedias set themselves up as meta-authorities more trustworthy than the sources they monitor.

The prophecy that Wikipedia will survive as one of a wide array of encyclopedias and metapedias is complicated by the fact that Wikipedia is already a metapedia whose every revision seeks to correct its earlier failings. Indeed one might argue that all encyclopedias are metapedias that aim, like Wikipedia, to summarize consensual scholarship rather than creating new theories, disseminating rather than generating information, with every new edition of *Britannica* a metapedia that seeks to update, supplement, and correct the immediately preceding edition. If Wikipedia survives as one of many encyclopedias, it is more likely to be noteworthy for its size than for any distinctive policies, procedures, or attitudes toward its own or others' authority.

Wikipedia, the Game

Except for the tragedy of the commons that Hardin proposes, these prophecies all assume the survival of Wikipedia, its online counterparts, and something like the Web itself in very much their current form. Other, murkier futures would follow from prophecies of more fundamental change. Its accommodation of playful impulses like those outlined in the previous chapter, for example, could end by turning Wikipedia into a full-blown interactive game. Phil Shuman, reporting for Fox News 11 in Los Angeles on 26 July 2007 on the systematic harassment of Web users, introduced his story by saying, "They call themselves Anonymous. They are hackers on steroids, treating the Web like a real-life video game."[32]

Shuman assumes that the inheritors of the Anonymous mantel could be moved only by negative motivations like bullying, revenge, or vandalism. But Wikipedia, for example, might easily be seen as a video game in the more even-handed way proposed by James Paul Gee in *What Video Games*

Have to Teach Us about Learning and Literacy.[33] Steven Johnson has argued that, if electronic media had emerged before novels, users might well dismiss the interloping new medium as single-modal, socially isolating, unavailable to the illiterate and dyslexic, and so hopelessly linear and noninteractive that "you simply sit back and have the story dictated to you" by a medium that "instills a general passivity" in its users.[34] Gee pushes this argument still further, countering the prejudice against video games by users predisposed to older media by enumerating no less than thirty-six learning principles fostered by different video games. His fundamental argument is that each game constitutes a semiotic domain—a coherent, self-contained world whose every element signifies something meaningful—governed by a design grammar players must intuit to navigate like experimental scientists. The skills required to master video games only begin with the principles of active critical learning, design, semiotic, semiotic domains, and metalevel thinking about semiotic domains.[35] Reconfiguring Wikipedia as an online game would reveal an analogous set of principles necessary for success: knowing one's audience, disagreeing diplomatically, resolving disputes through persuasion or transcendence, managing time efficiently, choosing one's battles, reading with an eye toward separating the wheat from the chaff. Like the new modes of online literacy Howard Rheingold has urged web users to cultivate, these habits are far from the modes of literacy engaged by *Britannica,* but they are no less valuable for being different.

Wikipedia and Web 3.0

The future of Wikipedia, which is widely seen as a quintessential avatar of Web 2.0, would be very different if it were overtaken by the emergence of Web 3.0, whose vision is commonly traced back to a prophecy Web inventor Tim Berners-Lee made in 1999: "I have a dream" in which

> the Web becomes a much more powerful means for collaboration between people [and] collaborations extend to computers. Machines become capable of analyzing all the data on the Web—the content, links, and transactions between people and computers. A "Semantic Web," which should make this possible, has yet to emerge, but when it does, the day-to-day mechanisms of trade, bureaucracy and our daily lives will be handled by machines talking to machines, leaving humans to provide the inspiration and intuition. The "intelligent agents" people have touted for ages will finally materialize.[36]

An online world in which Web pages are designed for computers, not humans, to read and people have no need to take a direct role in the innumerable routine transactions of commerce, citizenship, or information sharing will doubtless seem improbably utopian to some citizens and improbably nightmarish to others, particularly those whose ideas about computers programmed to take the initiative have been driven by science-fiction fantasies from Isaac Asimov's *I, Robot* to the *Terminator* films to Spike Jonze's *Her*. But we approach this scenario more closely every time we use a GPS to create maps and provide turn-by-turn instructions or call on a goddess in the smartphone like Siri. It is only a matter of time before scientists devise ways to connect GPS devices more directly to automotive engines or allow personal assistants into make preemptive plans based on their data-driven analyses of users' needs and cut the human role in making these decisions to the vanishing point.

The Wikimedia Foundation has taken a step into the world of Web 3.0 with Wikidata, "a free *knowledge base* that can be read and edited by *humans* and machines alike."[37] Wikidata is designed to provide a more centralized and highly structured repository of information for all the languages used by members of the Wikipedia community and their computers. It began as a list of links to all the Wikipedia entries in different languages on a given subject, but many of its entries have grown to incorporate links to corresponding entries in Wikipedia, Commons, and sources outside Wikimedia. Wikidata's next planned stage, which will probably be well under way by the time this book is published, will enable the automatic translation and updating of articles consisting of lists (e.g., United States cities by population) that are long, factual, and subject to frequent change.

Despite the Wikimedia Foundation's investment in Wikidata, whatever role Wikipedia plays in Web 3.0 is bound to be dramatically diminished from its paradigmatic role in Web 2.0. Wikipedia's hallmark has always been its openness to human interaction and human community. Its most likely evolution in Web 3.0, already presaged by Wikidata and the bots that flag new edits for examination, is into the encyclopedia that any computer can edit. The tirelessness of computers might seem to predict a future of unending editing wars. But it seems more likely that the machines, which have the reputation of being more logical than their programmers, would be more inclined to rise above unresolved partisan conflict—to transcend it, in Kenneth Burke's terms, by discounting the participants' motives rather than taking sides. In retrospect, the roistering online communities nurtured by Web 2.0 would appear as a transitional phase between the heroic, individualistic view of human knowledge, creativity, and accomplishment fostered by the Renaissance and sharpened by Enlightenment and Romantic aesthetics and the low-friction, maximally efficient information exchanges of Web 3.0.

The rapid rise and acceptance of Wikidata, editing bots, GPS, and Siri are only the most obvious signs that this particular future is already upon us. Cars that already park themselves will begin to drive themselves, reducing traffic accidents and fatalities as they turn driving time into leisure time. The clearest harbinger of this revolution in Web design is the radical downsizing of computer hardware. The UNIVAC computer of the 1950s took up an entire room. The personal computer of the 1980s sat on a desktop. The laptops of the 1990s could be carried through airports and into business meetings and coffee shops. But the generation currently coming of age online increasingly uses tablets and smartphones rather than dedicated computers to access the Web. For them, a computer is a tool that fits into a pocket.

The ability to make computers ever smaller every few years was predicted nearly fifty years ago by Moore's Law, named for Intel cofounder Gordon E. Moore. It asserted that the power that can be built into a state-of-the-art computer, based on the number of transistors on integrated circuits, doubles every two years. As circuit boards become tinier and tinier without any loss in computing power, computer hardware can shrink along with them. This technological curve is supplemented by a sociological development. Online subscribers are increasingly choosing the universal access to cyberspace represented by the virtual keyboards of iPhones over the inputting ease represented by desktop computer keyboards or even the physical keyboards of Blackberries. Young people have mastered the skills of rapid two-finger typing on phone keyboards of only a few square inches because they value connectivity over the accuracy, subtlety, or length of their queries and communications. Their lack of concern is both explained and cultivated by interactive search engines that begin to make educated guesses about what their users are looking for after they have typed a single character into the query box. These interactive search engines (along with Siri, GPS, and editing bots) are the advance guard of Web 3.0.

When cyberspace is a domain whose principal inhabitants are computers talking to other computers, Wikipedia's function will be reduced from a forum for communal debate and consensus about what is true and important and why to a provider of information required instantly and in small, easily digestible bits. Where is the nearest gas station? Does the Italian restaurant three blocks away take credit cards? How late do local drugstores stay open? When do flights leave New York airports for Berlin? Queries like these, so well suited to smartphones, are unlikely to be directed to Wikipedia. Not only is Wikipedia, whose rate of new articles and new edits has been slowing ever since it reached a peak in 2007,[38] ever likely to grow so detailed or local in the information it provides, but smartphone queries are increasingly likely to involve dedicated apps that bypass the Web. And these queries depend on literacy skills that are far more pragmatic and circumstantial than the more critical and discursive skills fostered by Wikipedia. Even though Wikipedia entries that editors deem too long are constantly being split, a smartphone will never be the optimal hardware for reading the Wikipedia article on Barack Obama Citizenship Conspiracy Theories or for scrolling through the five hundred footnotes to the article on Taylor Swift. Academics who fret over the marginalizing of liberal education as increasingly insular and irrelevant can at least take comfort from the likelihood that Wikipedia will also play a decidedly more marginal role for computer users who carry their hardware with them literally everywhere they go.

It does not follow from this that smartphone users are uninterested in interactivity. As all the world knows, social media like Facebook have joined hardware like the iPhone and communications technologies like Twitter to promote incessant networking. On the whole, however, this networking is elective, decentralized, and critical only in the connective choices it makes, not the way it evaluates them. Many blogs and information sites, for example, follow Facebook in allowing readers to press a button indicating that they like a given contribution but not that they dislike it. Liking costs nothing, as it is only inferentially comparative; nothing prevents a given reader from liking every possible response to a forum on a knotty question like academic freedom or gun control.

Jack Dorsey, the cofounder of Twitter, has traced its roots to his desire to free Instant Messaging from computer terminals and its name to a dictionary neighbor of "twitch" whose definition was "a short burst of inconsequential information." As Dorsey explains, "bird chirps sound meaningless to us, but meaning is applied by other birds. The same is true of Twitter: a lot of messages can be seen as completely useless and meaningless, but it's entirely dependent on the recipient."[39] Unlike Wikipedia, which painstakingly preserves the history of every entry it includes, and Deletionpedia, which archives entries even Wikipedia has declined to preserve, Twitter has 500 million users who glory in its evanescence. Restricting each tweet to 140 characters promotes immediacy over resonance and staying power and guarantees a feedback loop in which tweets demand ever more tweets in response. The photo- and text-sharing application Snapchat offers a deeply equivocal view of the next logical step. Perhaps its most distinctive advantage is the planned evanescence of every entry, which is viewable for only a short time before it disappears. At the same time, its terms of use warn that, although “you retain all ownership rights in your User Content,” it is also true that “by submitting User Content to Snapchat, you hereby grant us a nonexclusive, worldwide, royalty-free, sublicensable and transferable license to use, reproduce, modify, adapt, publish, create derivative works from, distribute, perform and display such User Content in connection with the Services.”[40] Everything posted to Snapchat is guaranteed to disappear down the memory hole, unless it suits the owners to preserve it indefinitely.

Facebook’s subscriber list numbers over one billion, over half of them accessing the site through mobile devices.[41] More than any other technology, Facebook indicates a new crisis in authority when desktops designed for Web 2.0 are replaced by smartphones designed to anticipate Web 3.0. From a typical user’s point of view, questions of authority are either reduced to triviality (what does it matter whether your friend Sandy’s latest grilled cheese sandwich really was the best ever?), rendered purely local and pragmatic (Apple swiftly replaced AppleMaps with GoogleMaps on its iPhone 5 when AppleMaps’ many glitches made its maps and directions dramatically less reliable), or resolved by appeals to popularity rather than sources of presumed wisdom or truth (so that Pope Benedict XVI was powerful not because any ex cathedra pronouncement he might make about faith or morals was protected from error but because he had a prodigious number of Twitter followers). From a marketer’s point of view, however, the authority of Facebook is immense, for it has the ability to deliver vast amounts of extremely specific self-reported information about millions of users that make it possible to tailor advertising to them on a virtually individual basis and to track their travels and purchases in both cyberspace and the physical world. As the wikis of Web 2.0 are replaced by the social and technological interfaces of Web 3.0, authority will become more centralized and users, despite the dream of Tim Berners-Lee, may well be relegated to the status of suppliers of personal information and consumers of public information.

Wikipedia and the Endless Feedback Loop

Peering even farther into the future reveals dim outlines of a synergistically information-saturated landscape in which every source is constantly updated after consulting every other source, so that the meaning of “source” becomes as thoroughly deconstructed as Jacques Derrida famously claimed has always been the case when he asserted that “the entire history of the concept of structure . . . must be thought of as a series of substitutions of center for center."[42] In this post-Web world, the authority of Wikipedia is of a piece with that of every other resource, since they all draw incessantly on each other in an endless feedback loop. Mathew Ingram offers a glimpse into this future that finds a vanishingly small difference between breaking news stories, constantly monitored, edited, and rewritten from one edition of a newspaper or television broadcast to the next, and Wikipedia articles, in which editors who claim ownership of stories forge inchoate observations into semiofficial narratives that assume authority as attacks on their central premises become more sporadic.[43] The Web has not disappeared or receded in this indeterminate future, but information that appears there is no more or less authoritative than information made available through television, telephones, public libraries, or public education. The open-source news site Reddit, whose users' votes about which news stories are most important determine how prominently they are displayed ("the hottest stories rise to the top, while cooler stories sink"[44]), already provides a glimpse of a future in which the line between public information and social networking vanishes altogether.

Two implications of this future are especially noteworthy. The first is the unlikely alliance such a post-Web landscape would forge between Wikipedia and liberal education. To be more precise, conceiving of information as an instantaneously and incessantly revised feedback loop would deal an even more mortal blow to the authority of liberal education than to Wikipedia. Although college classrooms depend on give-and-take as students and their teachers question their core values, the process depends on their sharing a set of values and establishing them as a core. The values of intellectual freedom and critical thinking may not be timeless, but their value depends on their enduring beyond the next news cycle. So an attack on these values that succeeded in annihilating them would mark both Wikipedia and liberal education as impossibly old-school forums, leaving them both in ruins.

The second implication also concerns Wikipedia's relationship to liberal education. The value Wikipedia has for liberal education today depends on the novelty that encourages its users to question assumptions about problems of authority in reference sources and liberal education. But it is far from clear what will become of that value for a generation of users who have grown up with Wikipedia and may not consider its authority, along with that of Web 2.0 or Web 3.0, novel enough to be worth examining critically at all. In addition to finding itself more closely aligned with liberal education, Wikipedia might well depend on liberal education to continue raising the questions that give it its potency.

When the Academy Bets on the Web

For their part, colleges and universities, initially skeptical of online resources, have begun to embrace the Web, a trend that promises to continue apace. Yet this embrace is highly selective, driven not only by the new research opportunities offered by cyberspace but by the economies forced on higher education by declining public support and a recession that shocked all but the most well-insulated endowments. The exodus of undergraduates from humanities courses to preprofessional programs that promise immediate practical value is accelerated by corporate donors who, no longer concerned to foster a managerial class whose background includes liberal education, focus their funding on programs more directly connected to their businesses. The management of American colleges has at the same time become widely corporatized, with decisions about institutions' missions and futures made by administrators trained in management rather than administrators risen from the professorial ranks. College students are increasingly treated as consumers of knowledge, a product to be supplied to them as efficiently as possible. All these changes motivate colleges to seek ever-greater economies in the delivery of instruction. Across the country, colleges are replacing research scholars with teachers who are expected to spend less time in libraries and laboratories so that they can spend more time in the classroom and preparing lessons for online classes. When planners at these colleges look to the Web, it is only natural for them to see it as another cost-cutting tool.

The interactive Web does indeed allow prodigious savings in many areas. Libraries can replace expensive subscriptions to print editions of journals with more economical online subscriptions and use services like Project Muse to give their community access to searchable databases and resources to which they do not subscribe. Newly minted professionals who would otherwise face daunting challenges in placing manuscripts with hard-pinched university presses can publish their work online in peer-reviewed digital venues that can bring state-of-the-art research to a much wider audience much more quickly than academic journals can. Instructors can bring a dizzying variety of materials into their classrooms though PowerPoint displays or online connections. Initiatives in distance education draw online students whose work schedules or geographic distance would prevent them from taking courses on campus.

The most recent development in online education is the spread of massive open online courses, known as MOOCs, a logical extension of distance education. An instructor who offers a MOOC has the potential to reach thousands of students, most of whom have not paid for the course and do not participate actively in online discussions. Students' performance can be evaluated either by objective quizzes and exams that can be graded by computers or by peer review and other modes of crowdsourcing. Even if only a small fraction of the students enrolled in a MOOC pay tuition in order to receive formal course credit or some other formal recognition that they have completed the course, the potential savings over typical classroom models of instruction are enormous. So it is no surprise that the best-known players in this area are for-profit companies like Udacity and Coursera.

All in all, then, college education in the United States is trending away from one-size-fits-all programs like the Great Books curriculum born at the University of Chicago and toward cafeteria or buffet curricula—individualized, self-selected, with fewer requirements—and technology-driven modes of interaction. Although professors may miss the give-and-take typical of classroom discussions, there is no obvious sign that students, brought up on smartphones and social media and identifying their main reasons for attending college as earning economically valuable credentials and enjoying an undergraduate lifestyle, do so as well. The experiences contemporary students seek from such interactive experiences are far more elective, decentralized, and self-chosen than submission to a series of leading seminar questions, some of which may lead to answers, some to further questions.

The Future of Play

Many sketches of Wikipedia's future, like those of Web 3.0, MOOCs, or liberal education generally, are based on economic projections. The most frequent narrative offered as an alternative to these prophecies is some version of Yochai Benkler's vision of dawning wisdom and shared governance in which people avert the tragedy of the commons by rising above their individual selfish interests in the name of a larger, wiser community. In dubbing economics the dismal science, the historian Thomas Carlyle meant both to contrast it with what he called the gay science involved in writing poetry and to mock the doom-laden prophecies of Thomas Malthus, who predicted that the world's population would grow so much faster than its ability to produce food that mass starvation would be the inevitable result. Garrett Hardin, noting how much more difficult it is to legislate temperance than prohibition, contends that "the great challenge facing us now is to invent the corrective feedbacks that are needed to keep the custodians [of the law] honest. We must find ways to legitimate the needed authority of both the custodians and the corrective feedbacks."[45]

But it is possible to imagine an alternative to economically determined futures for Wikipedia without resorting to utopian fantasies or legal sanctions by emphasizing the playfulness of the enterprise. Wikipedia's future may be determined by economics, but its contributors are motivated by something else: the pursuit of intellectual capital or expert capital or community capital but certainly not the pursuit of money. This is not to say that playful motives are incompatible with economic motives, only that we need to examine the relation between the two more closely. Recent developments in liberal education, traditionally a seriously playful enterprise, allow us to do just that.

Many humanities scholars, not content to join their administrations in seeking more economically efficient channels through which to deliver instruction, have sought other ways to address the challenges and opportunities online culture offers. Some of them have adopted the term "digital humanities," often abbreviated DH, to describe the common grounds that underlie their diverse projects. DH researcher Patrik Svensson offers a usefully broad definition when he says that "minimally, digital humanities is manifested by a single scholar, teacher, artist, programmer, engineer or student doing some kind of work—thinking, reflecting, writing, creating—at the intersection of the humanities and information technology—or by 'products' resulting from such activities."[46] Noting the apt vagueness of this description, Svensson cites a DH typology Tara McPherson has proposed "that makes distinctions between the computing humanities, blogging humanities and multimodal humanities. According to McPherson, the computing humanities focus on building tools, infrastructure, standards and collections whereas the blogging humanities are concerned with the production of networked media and peer-to-peer writing. The multimodal humanities bring together scholarly tools, databases, networked writing and peer-to-peer commentary while also leveraging the potential of the visual and aural media that are part of contemporary life."[47]

The big-tent label of digital humanities might seem to focus on finding new ways digital landscapes and applications can be used for traditional instruction in the humanities in order to shrink its costs. Svensson acknowledges that "the individualized forms digital humanist and digital humanists are more commonly used in relation to the digital as tool (and the humanities computing tradition) than the digital as study object," but adds that "the landscape is shifting." He excerpts an e-mail conversation in which he remarks to media studies scholar Charles Ess, "My sense of internet studies (IS) is that it largely focuses on internet as a study object," and Ess replies, "I would say more on the sorts of human/social interactions that are facilitated by the technologies and applications."[48]

Matthew G. Kirschenbaum, agreeing that digital humanities is "a culture that values collaboration, openness, nonhierarchical relations, and agility," observes that the label has become something of "a free-floating signifier, one that increasingly serves to focus the anxiety and even outrage of individual scholars over their own lack of agency amid the turmoil in their institutions and profession."[49] His analysis raises the inevitable question of whether digital humanities marks a bold new frontier for liberal education or its last gasp before it succumbs to a populist, decentered Web 2.0, a posthuman Web 3.0, and a series of economic reversals that have left liberal education ever more vulnerable to ideological attacks on its value in a world of digital communications. Digital humanities can readily be envisioned as a profit center for cash-strapped universities or a critical locus that encourages scholars and students to question and reformulate the foundational principles of liberal education.

Digital humanists have by and large neglected Wikipedia to focus on a future constantly experienced as just over the horizon. Typical is David Silver's definition of critical cyberbculture studies as

> a critical approach to new media and the contexts that shape and inform them. Its focus is not merely the Internet and the Web, but, rather, all forms of networked media and culture that surround us today, not to mention those that will surround us tomorrow. Like cultural studies, critical cyberculture studies strives to locate its object of study within various overlapping contexts, including capitalism, consumerism and commodification, cultural difference, and the militarization of everyday life.[50]

Yet what Henry Jenkins calls Wikipedia's "moral economy of information"[51] and Joseph Michael Reagle Jr. calls the "*good-faith* collaborative culture of Wikipedia"[52] is likely to continue to provide a compelling laboratory in which claims about collaboration, scholarship, and authority can be playfully, seriously tested.

Three Tomorrows for Authority

Whatever may become of Wikipedia—or of its place in the future of the Web or digital humanities or liberal education—its array of possible futures all reveal one of three attitudes toward authority.

The first of these is absolute allegiance toward those authorities who demand uncritical acceptance of their claims. Fox News, MSNBC, and the editorial pages of the *Wall Street Journal* acknowledge competing authorities only to deprecate them. Such authoritarian sources might seem tyrannical if it were not so clear that large numbers of consumers deliberately seek them out. Bill Bishop has traced the processes that have encouraged Americans to associate more closely only with fellow citizens who share their habits, assumptions, and beliefs and the implications of living in a culture that is better described as a collection of tribes or microcultures, each one willfully deaf to the claims of the others. This process, which Bishop calls "the big sort," has a serious downside: "As people seek out the social settings they prefer—as they choose the group that makes them feel the most comfortable—the nation grows more politically segregated—and the benefit that ought to come from having a variety of opinions is lost to the righteousness that is the special province of homogeneous groups."[53] From its beginnings, the Web has fostered this process of self-balkanization, allowing online browsers ever more efficient ways to surround themselves with opinions and points of view exactly like their own without taking the trouble to relocate their households physically. The radically elective communities favored by cellphone users who neglect their dinner companions in order to respond to the texts of their online friends take this development to its logical extreme, as individuals embrace virtual communities of the like-minded while ignoring their neighbors in what comes to seem more and more arbitrary physical proximity. Online neighborhoods that foster this closed sense of community and the absolute authority it breeds include Conservapedia, Encyclopedia Dramatica, and any number of sites denying the Holocaust or demanding gun-control legislation. Unwary students who enter any of these sites are in effect wandering into a self-enclosed universe whose claims brook neither disagreement nor critical examination. To take a particularly flagrant example, none of the hotlinks ("Historical Writings: Essays, Sermons, Speeches & More"; "The King Holiday: Bringing the Dream to Life"; "Civil Rights Library: History of People and Events"; "Attention Students: Try Our MLK Pop Quiz"; "Learn more about Kwanzaa!!") displayed on the innocuous-sounding home page martinlutherking.org, for example, indicates just how tendentious the site is. Only following those links reveals—and indeed might not reveal to every visitor—that it is a white supremacist site devoted to vilifying King.

The proliferation of such sites suggests a second future for communal attitudes toward authority. Instead of declaring allegiance to a single central authority like *Britannica* or Wikipedia, citizens may parcel out their allegiance to a wide range of specialized authorities like the Animal Diversity Web and Rocklopedia Fakebandia that do not compete with each other because their claims never overlap. Hence Andrew Keen's prediction that "Silicon Valley is going to rediscover expertise. It's going to be the new new thing. The cult of the amateur is going to be replaced by the cult of the expert."[54]

Numerous objections could be lodged against this prediction. The future it adumbrates sounds distinctly retro, a vision that recalls Franklin Roosevelt's brain trust and the Ivy League of the 1950s. It is not clear that Silicon Valley has so far failed to respect experts or that the most likely futures for attitudes toward authority are really circumscribed by two such polarized visions as Keen's cults of amateurs and experts. As Clay Shirky has noted, "Wikipedia is the product not of collectivism but of unending argumentation; the corpus grows not from harmonious thought but from constant scrutiny and emendation."[55] In one respect, however, Keen's prediction deserves serious consideration. Even if respect for experts does not normally take the form of cults, it cannot help encouraging the formation of tribes or clans. The authority of specialized experts, as Mathieu O'Neil observes, is not decentralized; it is that of an "online tribal bureaucracy."[56] Citizens who recognize the authority of multiple experts are simply declaring their allegiance to multiple—and, if they are lucky, noncompeting—tribes. Since few people are aware that they may require wisdom about the ideology of class warfare, personal retirement strategies, and the simplest ways to patch holes in wallboard all on the same day, these allegiances are normally experienced sequentially rather than simultaneously. But that does not make them any less exclusive, clannish, or tribal. The implied conflicts among ostensibly noncompeting authorities often take the form of caste distinctions between such bodies as research universities and community colleges or tenure-track faculty members and full-time adjunct faculty members, each with its own distinctive mission but with some missions tacitly assumed to be more equal than others. The fact that the dominant tribe's claims to authority are assumed rather than explicitly argued does not make them any less absolute.

Readers who find the two alternative futures represented by allegiance to an unquestioned, authoritarian central authority and allegiance to a series of tribal authorities unappetizing, and perhaps not all that much different from each other, are most likely to be interested in a third future whose defining feature is not the authority claimed by specific sources but the authority conferred on those sources by their followers: a contingent, ad hoc authority that is constantly discounted even by its most devoted followers, who retain a keenly critical eye every time they consider the motives, claims, and limits of the authorities they hold most dear.

Facebook users running graph searches of other Facebook users are seeking information or resources, from people who like both Mozart and Nine Inch Nails to gay attorneys in Cleveland. But people who talk to their friends about their problems, either online or offline, are much less likely to be so directive. Sometimes they seek particular advice, sometimes opinions, sometimes only a sympathetic ear. In the same way, citizens and researchers can seek and acknowledge conflicting authorities without simply surrendering to cynicism because they sometimes fail to yield consensus. A common complaint grade school teachers make about their students, for example, is that when they search for answers to their questions online, they assume that the sources they find will be absolutely authoritative without taking the trouble to confirm that authority. These students do not recognize that such an unconditional acceptance of authority is more likely the product of ignorance or incomplete knowledge than the more rounded perspectives encouraged by discounting. Nearly every entry on Wikipedia, for better or worse, claims unconditional authority about its subject on its main page at the same time that it displays some of the ways those claims have been discounted on its Talk and History pages. Perusing these pages and considering their relations to each other amount to an education in critical discounting.

Teachers seeking to model more critical attitudes toward authority for their students are well aware that such attitudes are not something anyone slides into naturally. Some people are born naturally credulous and others naturally skeptical, but no one is born intelligently critical. Most people crave authority and are drawn to it—even the mantra "Question Authority" asserts a bumper sticker authority of its own—but no one spontaneously begins to discount authorities unless they compete directly under circumstances that invite observers to analyze and transcend the conflicts between them. The tendency of any system is toward a rising entropy that favors the uncritical acceptance of authority in both online and collegiate communities. Teachers who want to foster more widely critical attitudes toward authority must work toward this goal and motivate their students to work toward it as well. Using Wikipedia thoughtfully and discriminatingly has a vital role to play in this process. So does liberal education itself, whose future, after all, is no more certain than that of Wikipedia.

To discount authority is neither to accept it unquestioningly nor to ignore it completely but rather to realize that it is constantly in play. Citizens who regard authority playfully recognize that every authority is contingent, hypothetical, and negotiable because it competes with other authorities. Every authority carries within it potentially authoritarian claims, but its followers need not accept these claims at face value. Indeed, the better educated and more experienced they are, the more likely they will be to treat these claims as always in flux, always up for grabs.

In announcing his belief that "there will *always* be a base for someone to move forward on,"[57] Jimmy Wales acknowledges that Wikipedia, which seeks to make its every entry as authoritative as possible, is always oriented toward a time in which its present authority will be judged, perhaps harshly, in the light of its future, presumably more authoritative authority. As a resource that incorporates and discounts the ideals of both perfectibility and play, Wikipedia reveals with startling clarity the paradoxes inherent in the notion of authority and its proper place in the social world. Very few forums simultaneously promote their own authority and take an active role in teaching their participants to question authority. In doing so, Wikipedia, like the whole project of liberal education, offers an exceptionally useful platform for moving from teaching how to follow authority to teaching about different kinds of authority and from there to teaching how to evaluate authority and ultimately to teaching how to develop and use one's own authority.

When he was interviewed in *Truth in Numbers,* Web analyst and developer Mark Pesce observed that "what we're starting to move into is a world where what the truth is becomes a question of how you trust."[58] Studying Wikipedia reminds us that we have always lived in such a world. The habit of considering critically whom you trust and why may have been repressed by print culture with its emphasis on rote memory and adherence to authority. But this habit is preserved and constantly reawakened by liberal education. As Yochai Benkler puts it: "the emergence of a substantial sector of nonmarket production, and of peer production, or the emergence of individuals acting cooperatively as a major new source of defining widely transmissible statements and conversations about the meaning of the culture we share, makes culture substantially more transparent and available for reflection, and therefore for revision."[59]

Playing with culture makes it more transparent and subject to revision, whether the players are editing Wikipedia, questioning the foundations of liberal education, or participating in their own governance. Howard Rheingold has cited numerous examples of "wiki government"[60] that seek to give citizens a more direct role in formulating policy. If governing bodies can acknowledge the importance of more widespread public participation in making and reviewing legal policies, surely the time has come for college teachers to acknowledge that whenever liberal education succeeds, it is because it teaches its students to play responsibly with authority: to recognize and negotiate between conflicting authorities, to accept authority while still discounting it, and ultimately to stake claims to the authority they deserve while recognizing that many competitors deserve it as well.

APPENDIX

Exercises for Exploring Wikipedia and Authority

Traditional philosophical analyses of authority seek through categorical organization and logical argument to contain and resolve the problems and paradoxes they raise. Although I hope this book is logical and well-organized, it does not share this ultimate goal. Perhaps because I am not a professional philosopher, I am more interested in flushing out problems and paradoxes of authority than in resolving them. Instead, I encourage readers to ponder them and come up with their own responses, which may or may not take the form of resolutions.

In pursuit of that goal, I have outlined fifty-one exercises mostly framed as classroom activities. Most of them are addressed to students in hypothetical courses in composition or research methods. But others are addressed to their teachers, still others to colleagues or companions or like-minded citizens or myself. I offer them frankly as thought experiments for anyone reading *Wikipedia U.* They are designed to stimulate discussion, but the discussion can as readily be interior as exterior, and one sort of discussion they aim to stimulate is over the question of whether better exercises could be designed.

This book would remain incomplete if no one ever argued with it. The exercises I have provided offer a few ways to open that argument, my own humble version of the crowdsourcing of which Wikipedia is such a stellar example. They have already provided me with many opportunities to argue with myself, and I hope they will provide readers with many more opportunities of their own.

Introduction: The Battle of the Books

Exercise 1. Find a guide to online research published before the founding of Wikipedia—for example, Eric Crump and Nick Carbone, *English Online: A Student's Guide to the Internet and the World Wide Web* (Boston: Houghton Mifflin, 1997); Andrew Harnack and Eugene Kleppinger, *Online! A Reference Guide to Using Internet Sources* (Boston: Bedford/St. Martin's, 2000); the first edition of William B. Badke, *Research Strategies: Finding Your Way through the Information Fog* (San Jose: Writers Club, 2000); or any of the first three editions of *Manual of Online Searching Strategies,* edited by C. J. Armstrong and Andrew Large (Burlington: Gower, 1988, 1992, 2001). Using the information and advice in your guide, write two accounts of how you would look for information about the Great Wall of China online in 2000 and now.

Exercise 2. On whose shoulders do you stand? Make a list of half a dozen people who have been most important in shaping your intellectual habits and specify what legacy each one of them has left you.

Chapter One: Origin Stories

Exercise 3. Redesign Wikipedia's opening page and explain why you made the decisions you did about its new look.

Exercise 4. List three things that tend to get better over time and three things that tend to get worse. Is Wikipedia more like the items on the first list or the second?

Exercise 5. Find a wiki or blog devoted to discussing liberal education, for example, the one hosted by *The Chronicle of Higher Education* (http://chronicle.com/section/Blogs/164). Follow at least one thread of it for a week, contribute to it at least twice, and report back summarizing your experience as both reader and contributor and explaining how the wiki or blog's mission and method are, or are not, compatible with its assumptions about liberal education.

Exercise 6. List the qualities you would most greatly value in a new computer, in a new reference source, and in a college you are attending or would like to attend. Then write a brief essay comparing the three lists.

Exercise 7. Use Google, Bing, Yahoo, or some other search engine to look up several topics until you find one whose Wikipedia entry is not the first item the search engine lists. Then compare the first-listed item and the relevant Wikipedia entry and make your best case for which one should be listed first and why.

Exercise 8. Look through a recent issue of a scholarly journal and a recent collection of essays in your academic field, make a note of the date of the sources each reference cites, graph the results, and discuss any patterns that emerge.

Chapter Two: Paradoxes of Authority

Exercise 9. Draw a chart illustrating Wikipedia's administrative structure, compare it with the administrative structure of your college or the U.S. federal government, and indicate how it could be improved.

Exercise 10. Review http://en.wikipedia.org/wiki/Wikipedia:Version_1.0_Editorial_Team/Assessment; read the examples it gives of the Wikipedia articles labeled A-Class, B-Class, and C-Class; describe the differences among them that indicate why they have been rated differently; and indicate whether you accept the ratings, which may be several years out of date, and what you could do to help promote the B-Class and C-Class articles to the next stage.

Exercise 11. Choose one of the myths or legends catalogued at http://snopes.com or some other site devoted to debunking legends, find at least three places on the Web where it is presented as fact, and compare the evidence for and against its truthfulness.

Exercise 12. Choose two students to play two hypothetical experts who conduct a debate on a given subject. Then choose two more students to play two hypothetical judges to establish and debate the credentials of the two experts. Then choose two more students to establish and debate the expertise of the two judges. Discuss the results of all three debates. Would further debates on the expertise or qualifications of the debaters clarify these results?

Exercise 13. List three hard-copy print sources from which it is permissible to quote material for academic papers, three more sources from which it is not, and three more from which it might be under certain circumstances. Then compare the three lists and their rationales with a view toward forming general rules about what academic papers can and cannot legitimately quote under what circumstances.

Exercise 14. Have everyone in the class write a brief paragraph on one of these topics—slavery, UFOs, trickle-down economics, Rev. Sun Myung Moon, or Planned Parenthood—from a neutral point of view. Drawing on as many of these paragraphs as necessary, compile a single paragraph that represents the class's best sense of a neutral discussion of the topic. Then discuss the problems you faced in compiling these consensual paragraphs and the possible challenges they might face from other quarters.

Exercise 15. Write or act out a debate about plagiarism and originality involving each of the following people:

- A researcher accused of plagiarism by a Wikipedia system operator, or sysop
- The author of the work the researcher is accused of plagiarized
- The reader who first brought the accusation
- A sysop who defends the researcher on the basis of NOR

Exercise 16. Make a list of all the people who could reasonably be counted as experts or concerned parties concerning the following proposition—"the Electoral College of the United States has outlived its usefulness and should be abolished"—and explain how you would weigh their respective credentials and claims to authority against each other.

Exercise 17. Wikipedia did not invent edit wars; it merely accelerated them and made them easier to discover and document. Choose one of the many scholarly disagreements in astronomy, biology, or psychology before the rise of Wikipedia and compare its course with the course of a single edit war of your choice.

Exercise 18. Imagine that you were an academic who had just won the Nobel Prize in Chemistry, Physics, or Economics. How would your award change the way you taught your introductory undergraduate course?

Chapter Three: The Case against Wikipedia

Exercise 19. Discuss which of the following groups has done most to promote human knowledge. Divide the class into five groups and organize a debate in which each group defends the claims of one of the following parties:

- IBM
- Apple
- Microsoft
- Google
- Wikipedia

Exercise 20. Consult as many or as few online sources as you like about the correct spelling of "Sydney [or Sidney] Greenstreet," "Tacoma [or Takoma] Park," or the Ray Evans and Jay Livingston song "Que Sera, Sera" [or "Qué Sera, Sera," or "Que Será, Será"]. Then explain which spelling is correct and how you can tell.

Exercise 21. Explain whether it is better to resist or yield to seduction and why.

Exercise 22. Propose a mechanism for disconfirming errors in humanities research and publicizing these refutations and explain how it would work.

Exercise 23. Imagine that an academic had written to a scholarly press or a university hiring committee promising an entirely original approach to a specific intellectual problem or discipline or quest for knowledge and write a letter in response from either an editor at the press or the chair of the hiring committee implicitly indicating which kinds of originality the press or college valued and which kinds it stigmatized or avoided.

Exercise 24. Divide the class into pairs of students and have each student prepare a brief, neutral analysis of one of the following problems: the difficulties in maintaining the economic health of the European Union, the future of entitlement spending in the U.S., the advantages of print books versus e-books, the evidence for evolution or global warming. Ask students to exchange their outlines and have each point out places in the other's outline where neutrality breaks down. Then have each student read the Wikipedia page on Neutral point of view, invite each student to respond to the other's remarks by attempting to make the analysis more neutral according to Wikipedia's definition of NPOV, and discuss the results.

Exercise 25. Find three brief texts whose authors are unidentified and compare your sense of their authority to that of three brief texts on similar subjects whose authors are identified.

Exercise 26. List the courses you have taken so far in college and the courses you are likely to take in the foreseeable future and explain why you have taken or plan to take them.

Exercise 27. What sorts of professional expertise would you look for in each of the following people, and where would you look for evidence of that expertise?

- Someone to repair your leaking faucet
- Someone to tell you where to stay in a city you had never visited
- Someone to prepare your tax return
- Someone to advise you how to invest your money for long-range goals
- Someone to explain the differences among the world's leading religions

Exercise 28. If you were an editor in charge of assigning book reviews, which experts would you consider best qualified to review each of the following books:

- A study of women in the twenty-first-century American workplace (a well-known feminist? the CEO of a Fortune 500 corporation? an academic sociologist? a Frenchwoman long resident in America?)
- A new biography of Frank Sinatra (the author of a biography of Bing Crosby? the author of an earlier Sinatra biography? one of Sinatra's children? a contemporary pop singer?)
- A history of the papacy (a Roman Catholic priest? a prominent critic of the Church? an Italian journalist assigned to cover the Vatican? the author of a history of Italy?)

Exercise 29. Divide students into pairs and ask one student in each pair to write a memo from Jimmy Wales to all Wikipedia users establishing a maximum length for Wikipedia articles and explaining why and the other student to write a response protesting this new limitation.

Exercise 30. Write a brief proposal on one of the following topics:

- If you were allowed to teach the next meeting of one of your classes, what would you do?
- If you were allowed to redesign one of your courses, or design a new course from scratch, how would it look?
- If you could create the ideal college, how would it differ from your own college?

Chapter Four: Playing the Encyclopedia Game

Exercise 31. Make a list of five websites you consider trustworthy, reliable, or authoritative. Explain what makes each one trustworthy, what your trust is based on, and what kind of information you trust it to provide. Then compare each of your five explanations with the others and with explanations other members of your class have provided.

Exercise 32. Read the section "Approaches to Presenting Criticism" in the Wikipedia entry "Wikipedia: Criticism" (http://en.wikipedia.org/wiki/Wikipedia:Criticism). Compare the five strategies Wikipedia suggests for incorporating critical perspectives about the handling of a given topic—an integrated approach, separate "Reception" or "Criticism" sections, and separate "Reception of" or "Criticism of" entries—and explain which approach seems most even-handed and useful to you.

Exercise 33. Read the Wikipedia entry "Political Arguments of Gun Politics in the United States" (http://en.wikipedia.org/wiki/Political_arguments_of_gun_politics_in_the_United_States) and suggest three specific changes you would propose to the entry.

Exercise 34. Find a stub on Wikipedia, explain why it has remained a stub even though its topic has been identified as a candidate for a longer entry, and expand it.

Exercise 35. Choose a topic from Wikipedia's list of Requested Articles (http://en.wikipedia.org/wiki/Wikipedia:Requested_articles) and write a brief article on it.

Exercise 36. Compare the Talk pages for "Creationism," "George Soros," and "United States Fiscal Cliff." Then describe the range of models they provide that you might find useful to follow or avoid in editing Wikipedia pages.

Exercise 37. Write a response to a Wikipedia contributor who modified or reverted one of your own contributions to a Wikipedia entry and post it on that entry's Talk page. Indicate as judiciously as possible your reactions to the edit and your suggestions for continuing further dialogue.

Exercise 38. After reading and thoughtfully considering the Wikipedia page on "Authority," edit the page, alone or together with others, adding, deleting, or editing whatever seems most important, and then monitor it for two weeks and report the results of your changes.

Exercise 39. Browse Wikipedia's list of Protected Pages (http://en.wikipedia.org/wiki/Special:ProtectedPages), propose one new page to be protected from malicious editing, and explain why that particular page deserves protection.

Exercise 40. Read the Wikipedia page on the Quickpolls Rules (http://en.wikipedia.org/wiki/Wikipedia:Quickpolls_policy) that were established for a brief period in 2004. Then join three other students in debating the best ways to keep the damage done to Wikipedia by malicious users to a minimum.

Exercise 41. Develop a preliminary outline for a first-year online college course in methods of research that uses Wikipedia as a laboratory for research procedures. Your outline should consider the ways in which Wikipedia does and does not constitute a logical extension of the topics on which current online courses are based, its ability to draw on a preexisting audience and provide the broadest possible freedom in choosing research topics and the ultimate peer-editing experience, and the risks and limitations likely to arise in the course you have designed.

Chapter Five: Tomorrow and Tomorrow and Tomorrow

Exercise 42. Propose a new initiative for Wikipedia parallel to Wiktionary, Wikiquote, and Wikinews and explain how you would launch and market it in a way that is consistent with Wikipedia's stated strategic priorities.

Exercise 43. Write three advertisements keyed to particular Wikipedia pages and explain how you would want them placed on or linked to those pages and why.

Exercise 44. Read the Wikipedia entry "How to Contribute to Wikipedia Guidance" (http://en.wikipedia.org/wiki/Wikipedia:How_to_contribute_to_Wikipedia_guidance) and the "Proposals" section of the entry "Wikipedia: Policies and Guidelines (http://en.wikipedia.org/wiki/Wikipedia:PROPOSAL#Proposals). Then post a suggestion at the Policy or Proposals village pump, using the RfC (Request for Comments) tag, that edits be either more or less closely monitored and curated and report back on the responses over a period of a week. (Since a flood of such suggestions would be perceived as antisocial, this exercise is best carried out by a single party, or a group acting as one.)

Exercise 45. Write a proposal for a new metapedia, serious or playful, explaining how it would work, why it would be an improvement over specific encyclopedias currently available, and how it would be useful to a particular community.

Exercise 46. Write an analysis of Wikipedia as a game, find the best place to post it on Wikipedia, and report the responses over a week.

Exercise 47. Propose a set of software tools for a version of Web 3.0 that does not include robots.

Exercise 48. Write a story about a group of fugitives from Web 3.0 who turn outlaw, giving them ample opportunity to explain their principles.

Exercise 49. Brainstorm an online version of a course you have taken that would not require any hardware other than a smartphone.

Exercise 50. Make a list of the dozen websites you consider most authoritative, describe what you trust each one to be good at doing, and explain what it would take for a new or newly discovered site to crack your list.

Exercise 51. Imagine three distinct post-Wikipedia landscapes and explain which of them is most likely, which of them you prefer, and what you would be willing to do to work toward each of them.

NOTES

Introduction: The Battle of the Books

1. John Henry Newman, *The Idea of a University* (1852), edited by Frank M. Turner (New Haven: Yale University Press, 1996), p. 81.

2. *General Education in a Free Society: Report of the Harvard Committee* (Cambridge, MA: Harvard University Press, 1950), pp. 47, 46.

3. Jacques Barzun, *The House of Intellect* (New York: Harper, 1959), p. 252; see John Henry Newman, *An Essay in Aid of a Grammar of Assent* (New York, 1870), pp. 273–74.

4. Henry S. Broudy, *Truth and Credibility: The Citizen's Dilemma* (New York: Longman, 1981), p. 145.

5. Andrew Delbanco, *College: What It Was, Is, and Should Be* (Princeton: Princeton University Press, 2012), pp. 171–72.

6. Richard P. Keeling and Richard R. Hersh, *We're Losing Our Minds: Rethinking American Higher Education* (New York: Palgrave Macmillan, 2011), p. 43.

7. Louis Menand, *The Marketplace of Ideas,* American Council of Learned Societies Occasional Paper, no. 49 (n.p.: American Council of Learned Societies, 2001), p. 21.

8. Michael Bérubé, *What's Liberal about the Liberal Arts? Classroom Politics and "Bias" in Higher Education* (New York: Norton, 2006), p. 20.

9. Derek Bok, *Universities in the Marketplace* (Princeton: Princeton University Press, 2003), p. 30.

10. Alvin Kernan, *The Death of Literature* (New Haven: Yale University Press, 1990), p. 140.

11. Sven Birkerts, *The Gutenberg Elegies: The Fate of Reading in an Electronic Age* (New York: Faber and Faber, 1994), p. 188.

12. "Wikipedia, the People's Encyclopedia," *Los Angeles Times,* 13 January 2013, http://www.latimes.com/news/opinion/commentary/la-oe-gardner-wikipedia-20130113,0,5394695.story.

13. Cathy N. Davidson, "Humanities 2.0: Promise, Perils, Predictions," *PMLA* 123, no. 3 (2008): 707–17, at 711.

14. Kenneth Burke, *Attitudes toward History,* 3rd ed. (Berkeley: University of California Press, 1984), p. 244.

15. Burke, *Attitudes toward History,* p. 93.

16. Richard Arum and Josipa Roksa, *Academically Adrift: Limited Learning on College Campuses* (Chicago: University of Chicago Press, 2010), pp. 35, 122.

17. George D. Kuh, "What We're Learning about Student Engagement from NSSE: Benchmarks for Effective Educational Practices," *Change* 35, no. 2 (2003): 24–32, at 28, quoted in Arum and Roksa, p. 5.

18. Exodus 7:1–12.

19. 1 Kings 18:1–39.

20. Max Weber, *Economy and Society: An Outline of Interpretive Sociology,* translated by Ephraim Fischoff et al., edited by Guenther Roth and Claus Wittich (Berkeley: University of California Press, 1978), p. 213.

21. Weber, *Economy and Society,* p. 215.

22. Richard T. De George, *The Nature and Limits of Authority* (Lawrence: University Press of Kansas, 1985), p. 41.

23. E. D. Watt, *Authority* (London: Croom Helm, 1982), p. 106.

24. Richard Foley, "Egoism in Epistemology," in *Socializing Epistemology,* edited by Frederick Schmitt (Lanham: Rowman and Littlefield, 1994), p. 55.

25. Dennis H. Wrong, *Power: Its Forms, Bases, and Uses,* 3rd ed. (New Brunswick: Transaction, 2002), p. 35.

26. Weber, *Economy and Society,* pp. 999, 1001, 1002, 1001.

27. Evan Selinger and Robert P. Crease, Introduction to *The Philosophy of Expertise* (New York: Columbia University Press, 2006), p. 3.

28. Zalman Schachter-Shalomi and Natanel Miles-Yepez, *A Heart Afire: Stories and Teachings of the Early Hasidic Masters* (Philadelphia: Jewish Publication Society, 2009), p. 150.

29. Thomas Kuhn, *The Structure of Scientific Revolutions,* 2nd ed. (Chicago: University of Chicago Press, 1970), p. 10; Stanley Fish, *Is There a Text in This Class? The Authority of Interpretive Communities* (Cambridge: Harvard University Press, 1980), p. 304.

30. Robert K. Merton, *On the Shoulders of Giants: A Shandean Postscript,* 20th ed. (New York: Harcourt Brace Jovanovich, 1985), p. 45; see also pp. 267–69 and passim.

31. Quoted in Sam Williams, *Free as in Freedom (2.0): Richard Stallman and the Free Software Revolution,* 2nd ed., revised by Richard W. Stallman (Sebastopol, CA: O'Reilly and Associates, 2011), p. 155, http://static.fsf.org/nosvn/faif-2.0.pdf, consulted 29 July 2013.

Chapter One: Origin Stories

1. It is interesting to note that the ten Wikipedia portals listed around the puzzle globe are not those with the largest number of entries. The Dutch and Swedish Wikipedias, according to the list immediately below this graphic, both have more entries than the Polish, Japanese, Portuguese, or Chinese Wikipedias.

2. Andrew Lih, *The Wikipedia Revolution: How a Bunch of Nobodies Created the World's Greatest Encyclopedia* (New York: Hyperion, 2009), pp. 219, 220.

3. Ward Cunningham, "Invitation to the Patterns List," http://c2.com/cgi/wiki?InvitationToThePatternsList, consulted 29 July 2013.

4. Lih, *Wikipedia Revolution,* p. xv.

5. Lawrence Lessig, *Code and Other Laws of Cyberspace* (New York: Basic Books, 1999), p. 6.

6. Lih, *Wikipedia Revolution,* p. xv.

7. Lih, *Wikipedia Revolution,* pp. xv–xvi.

8. Lih, *Wikipedia Revolution,* p. xv.

9. Joseph Michael Reagle Jr., *Good Faith Collaboration: The Culture of Wikipedia* (Cambridge, MA: MIT Press, 2010), p. 1.

10. Lih, *Wikipedia Revolution,* pp. xxii, xvi.

11. Lih, *Wikipedia Revolution,* p. xviii.

12. Tim Berners-Lee, "Information Management: A Proposal," http://www.w3.org/History/1989/proposal.html, consulted 29 June 2013.

13. Tim Berners-Lee, "WorldWideWeb: Proposal for a Hypertext Project," http://www.w3.org/Proposal.html, accessed 29 July 2013; cf. Tim Berners-Lee, with Mark Fischetti, *Weaving the Web: The Original Design and Ultimate Destiny of the World Wide Web by Its Inventor* (San Francisco: HarperSanFrancisco, 1999), pp. 20–34.

14. Lih, *Wikipedia Revolution,* pp. xvii–xviii.

15. Eric A. Havelock, *Preface to Plato* (Cambridge, MA: Harvard University Press, 1963), pp. 300–301; Walter J. Ong, *Orality and Literacy: The Technologizing of the Word* (London: Methuen, 1982), pp. 49–53; James Gleick, *The Information: A History, a Theory, a Flood* (New York: Pantheon, 2011), pp. 35–41.

16. Reagle, *Good Faith Collaboration,* p. 19; compare Gleick, *Information,* pp. 382–83.

17. Andrew Brown, *A Brief History of Encyclopedias: From Pliny to Wikipedia* (London: Hesperus, 2011), pp. 15, 55, 105, 56, 87.

18. Brown, *Brief History of Encyclopedias,* pp. 88, 95, 96, 110.

19. Brown, *Brief History of Encyclopedias,* p. 84.

Chapter Two: Paradoxes of Authority

1. "Wikipedia: Why Wikipedia Is So Great," http://en.wikipedia.org/wiki/Wikipedia:Why_Wikipedia_is_so_great, consulted 30 July 2013.

2. "Wikipedia: Why Wikipedia Is Not So Great," http://en.wikipedia.org/wiki/Wikipedia:Why_Wikipedia_is_not_so_great, consulted 30 July 2013.

3. *Truth in Numbers? Everything, According to Wikipedia,* directed by Scott Glosserman and Nic Hill (Glen Echo Entertainment/Underdog Pictures, 2010), http://www.dailymotion.com/video/xqvyck_truth-in-numbers-everything-according-to-wikipedia_tech, consulted 30 July 2013.

4. "Wikipedia: About," http://en.wikipedia.org/wiki/Wikipedia:About, consulted 30 July 2013.

5. "Sysop Status," http://lists.wikimedia.org/pipermail/wikien-l/2003-February/001149.html, consulted 30 July 2013.

6. "Wikipedia: Becoming an Administrator," http://en.wikipedia.org/wiki/Wikipedia:Administrators#Becoming_an_administrator, consulted 30 July 2013.

7. "Wikipedia: Bureaucrats," http://en.wikipedia.org/wiki/Wikipedia:Bureaucrats, consulted 30 July 2013.

8. "Wikipedia: About," http://en.wikipedia.org/wiki/Wikipedia:About, consulted 30 July 2013.

9. "Stewards," http://meta.wikimedia.org/wiki/Stewards, consulted 30 July 2013.

10. "Board of Trustees," http://wikimediafoundation.org/wiki/Board_of_Trustees, consulted 30 July 2013.

11. "Bylaws," http://wikimediafoundation.org/wiki/Bylaws, consulted 30 July 2013.

12. Joseph Michael Reagle Jr., *Good Faith Collaboration: The Culture of Wikipedia* (Cambridge, MA: MIT Press, 2010), p. 133.

13. Mathieu O'Neil, *Cyberchiefs: Autonomy and Authority in Online Tribes* (London: Pluto, 2009), p. 172.

14. O'Neil, *Cyberchiefs,* p. 149.

15. O'Neil, *Cyberchiefs,* p. 158.

16. *The Truth According to Wikipedia,* directed by Ijsbrand van Veelen, *Tegenlicht,* 7 April 2008, http://www.youtube.com/watch?v=WMSinyx_Abo.

17. "Wikipedia: Featured Article Criteria," http://en.wikipedia.org/wiki/Wikipedia:Featured_article_criteria, consulted 30 July 2013.

18. "Wikipedia: Good Article Criteria," http://en.wikipedia.org/wiki/Wikipedia:Good_article_criteria, consulted 30 July 2013.

19. A complete list of the detailed criteria WikiProject Schools has established for articles about American primary and secondary schools is available at "WP: WikiProject Schools Assessment," http://en.wikipedia.org/wiki/Wikipedia:WikiProject_Schools/Assessment#Importance_scale, consulted 30 July 2013.

20. http://en.wikipedia.org/wiki/Main_Page, consulted 30 July 2013.

21. Susan Page, "Author Apologizes for Fake Wikipedia Biography," *USA Today,* 11 December 2005, http://www.usatoday.com/tech/news/2005-12-11-wikipedia-apology_x.htm, consulted 30 July 2013.

22. "CZ: Charter," http://en.citizendium.org/wiki/CZ:Charter, consulted 30 July 2013.

23. "Citizendium," http://rationalwiki.org/wiki/Citizendium#The_concept_of_expertise_on_Citizendium, consulted 30 July 2013. This entry includes a list of a dozen "pro-pseudoscience articles" on Citizendium.

24. "Wikipedia: Core Content Policies," http://en.wikipedia.org/wiki/Wikipedia:Core_content_policies, consulted 30 July 2013.

25. "Wikipedia: Verifiability," http://en.wikipedia.org/wiki/Wikipedia:V, consulted 30 July 2013.

26. "Wikipedia: Core Content Policies."

27. "Wikipedia: Core Content Policies."

28. "An Introduction to Everything2," http://everything2.com/title/An+Introduction+to+Everything2, consulted 30 July 2013.

29. Larry Sanger, "The Early History of Nupedia and Wikipedia: A Memoir," http://features.slashdot.org/story/05/04/18/164213/the-early-history-of-nupedia-and-wikipedia-a-memoir, consulted 30 July 2013.

30. "The Holocaust," http://en.wikipedia.org/w/index.php?title=The_Holocaust&oldid=256574, consulted 30 July 2013.

31. "Holocaust Denial," http://en.wikipedia.org/wiki/Holocaust_denial, consulted 30 July 2013.

32. "Wikipedia: No Original Research," http://en.wikipedia.org/wiki/Wikipedia:No_original_research, consulted 30 July 2013.

33. "Wikipedia: Researching with Wikipedia," http://en.wikipedia.org/wiki/Wikipedia:Researching_with_Wikipedia, consulted 30 July 2013.

34. "Wikipedia: No Original Research."

35. Jimmy Wales, "Original Research," http://lists.wikimedia.org/pipermail/wikien-l/2004-December/017557.html, consulted 30 July 2013.

36. Zoney, "Original Research," http://lists.wikimedia.org/pipermail/wikien-l/2004-December/017562.html, consulted 30 July 2013.

37. Mark Richards, "Original Research," http://lists.wikimedia.org/pipermail/wikien-l/2004-December/017566.html, consulted 30 July 2013.

38. Shane King, "Original Research," http://lists.wikimedia.org/pipermail/wikien-l/2004-December/017610.html, consulted 30 July 2013.

39. Wales, "Original Research."

40. Shane King, "NPOV and Credibility," http://lists.wikimedia.org/pipermail/wikien-l/2004-December/017619.html, consulted 30 July 2013.

41. Jay JG, "NPOV and Credibility," http://lists.wikimedia.org/pipermail/wikien-l/2004-December/017688.html, consulted 30 July 2013.

42. Jay JG, "NPOV and Credibility."

43. Mark Richards, "NPOV and Credibility," http://lists.wikimedia.org/pipermail/wikien-l/2004-December/017719.html, consulted 30 July 2013.

44. Ray Saintonge, "NPOV and Credibility," http://lists.wikimedia.org/pipermail/wikien-l/2004-December/017775.html, consulted 30 July 2013.

45. Stanley Fish, *Is There a Text in This Class? The Authority of Interpretive Communities* (Cambridge, MA: Harvard University Press, 1980), p. 321.

46. Wales, "NPOV and Credibility," http://lists.wikimedia.org/pipermail/wikien-l/2004-December/017629.html. Compare the more recent formulation at "Wikipedia: Neutral Point of View," http://en.wikipedia.org/wiki/Wikipedia:NPOV, both consulted 8 August 2012.

47. Richards, "NPOV and Credibility."

48. Joseph Michael Reagle Jr., "NPOV and Credibility," http://lists.wikimedia.org/pipermail/wikien-l/2004-December/017661.html, consulted 30 July 2013.

49. Joseph Michael Reagle Jr., "Open Communities, Media, Source, and Standards," http://reagle.org/joseph/blog/social/epistemological-authority.html, consulted 30 July 2013.

50. "Fox News," www.foxnews.com, consulted 30 July 2013.

51. See, for example, Thomas Van Parys, "Review: *Film Adaptation and Its Discontents: From Gone with the Wind to The Passion of the Christ*," *Image and Narrative* 8: 3 (2007), and Joyce Goggin, "Review: Simone Murray, *Expanding the Discipline: The Adaptation Industry*," *Adaptation* 6.1 (2013): 128–31.

52. See Robert M. Hutchins, "The Great Conversation," in *The Great Conversation: A Reader's Guide to Great Books of the Western World*, 2nd ed. (Chicago: Encyclopaedia Britannica, Inc., 1990), pp. 46–73.

53. Stanley Fish, "No Bias, No Merit: The Case against Blind Submission," *PMLA* 103 (Oct. 1988): 739–48, at 743.

54. Thomas Kuhn, *The Structure of Scientific Revolutions*, 2nd ed. (Chicago: University of Chicago Press, 1970), p. 10.

55. Kuhn, *Structure of Scientific Revolutions*, pp. 85, 79, 80.

56. For a particularly resourceful account of one teacher's approach to this problem, see "Race, Class, Gender," in Michael Bérubé, *What's Liberal about Liberal Education? Classroom Politics and "Bias" in Higher Education* (New York: Norton, 2006), pp. 141–205.

57. Kuhn, *Structure of Scientific Revolutions*, p. 80.

58. Christopher Jencks and David Riesman, *The Academic Revolution* (Garden City, NY: Doubleday, 1968); Menand, *The Marketplace of Ideas*, p. 9.

Chapter Three: The Case against Wikipedia

1. Sorin Cerin, *Wikipedia: Pseudo-Encyclopedia of the Lie, Censorship, and Misinformation* (n.p.: CreateSpace, 2011), p. 67. Although Wikipedia currently carries no entry on Cerin in either English or Romanian, there is an English-language entry on him in Metapedia (http://en.metapedia.org/wiki/Sorin_Cerin, consulted 30 July 2013).

2. Daniel Brandt, "Wikipedia Watch," http://www.wikipedia-watch.org/, consulted 30 July 2013.

3. Neil L. Waters, "Why You Can't Cite Wikipedia in My Class," *Communications of the ACM* 50, no. 9 (1 September 2007): 15–17, at 16. Wikipedia naturally maintains an entry on academic studies of Wikipedia, http://en.wikipedia.org/wiki/Wikipedia:Wikipedia_in_academic_studies.

4. James Gleick, *The Information: A History, A Theory, A Flood* (New York: Pantheon, 2011), pp. 380–81.

5. Darcy DiNucci, "Fragmented Future," http://darcyd.com/fragmented_future.pdf, consulted 30 July 2013.

6. Tim O'Reilly, "What Is Web 2.0: Design Patterns and Business Models for the Next Generation of Software," http://oreilly.com/web2/archive/what-is-web-20.html, consulted 30 July 2013.

7. Andrew Lih, "Wikipedia as Participatory Journalism: Reliable Sources? Metrics for Evaluating Collaborative Media as a News Resource," Paper for the Fifth International Symposium on Online Journalism, University of Texas at Austin (16–17 April 2004), http://www.ufrgs.br/limc/participativo/pdf/wikipedia.pdf.

8. See Charles Knight and Sam Pryke, "Wikipedia and the University, a Case Study," *Teaching in Higher Education* 17, no. 6 (December 2012): 649–59, at 654, 656.

9. Andrew Keen, "Web 2.0: The Second Generation of the Internet Has Arrived. It's Worse Than You Think," *Weekly Standard,* 14 February 2006, http://www.weeklystandard.com/Content/Public/Articles/000/000/006/714fjczq.asp.

10. Andrew Keen, *The Cult of the Amateur: How Today's Internet Is Killing Our Culture* (New York: Doubleday/Currency, 2007), p. 4.

11. *The Truth According to Wikipedia,* directed by Ijsbrand van Veelen, *Tegenlicht,* 7 April 2008, http://www.youtube.com/watch?v=WMSinyx_Abo.

12. Whether Sanger and Wales are better defined as cofounders of Wikipedia or as employee and employer has been the subject of endless debate, much of it fueled by the two principals.

13. Karl Marx, *The German Ideology,* in Karl Marx and Friedrich Engels, *Works* (New York: International Publishers, 1976), 5: 47.

14. Keen, *The Cult of the Amateur,* p. 205.

15. James M. Wells, *The Circle of Knowledge: Encyclopedias Past and Present* (Chicago: Newberry Library, 1968), p. 10.

16. "Reliability of Wikipedia," http://en.wikipedia.org/wiki/Reliability_of_Wikipedia, consulted 30 July 2013.

17. Jim Giles, "Special Report: Internet Encyclopedias Go Head to Head," *Nature* 438, no. 15 December 2005 (900–901).

18. "Fatally Flawed: Refuting the Recent Study on Encyclopedic Accuracy in the Journal *Nature,*" http://corporate.britannica.com/britannica_nature_response.pdf, consulted 30 July 2013.

19. Mathieu O'Neil, *Cyberchiefs: Autonomy and Authority in Online Tribes* (London: Pluto, 2009), p. 152.

20. "Encyclopaedia Britannica and Nature: A Response," http://www.nature.com/press_releases/Britannica_response.pdf, consulted 13 July 2012.

21. "Britannica Attacks . . . and We Respond," *Nature* 440 (30 March 2006): 582, http://www.nature.com/nature/journal/v440/n7084/full/440582b.html), consulted 30 July 2013.

22. Untitled editorial, *Nature,* http://www.nature.com/nature/britannica/eb_advert_response_final.pdf, consulted 30 July 2013.

23. Gleick, *Information,* p. 383.

24. "Fatally Flawed."

25. Andrew J. Flanagin and Miriam J. Metzger, "From *Encyclopaedia Britannica* to Wikipedia," *Information, Communication and Society* 14, no. 3 (2011): 355–74, at 369.

26. Robert Gorham Davis, "The Professors' Lie," *Columbia Forum,* n.s. 1 (Fall 1972): 6–15, at 11.

27. One recent study, for example, concludes that "a construct as complex as creativity will never be localized in the brain, be it the right hemisphere, anterior cingulate cortex, or other locus." See Rex E. Yung et al., "Biochemical Support for the 'Threshold' Theory of Creativity: A Magnetic Resonance Spectroscopy Study," *Journal of Neuroscience* 29, no. 16 (22 April 2009): 5319–25, at 5322.

28. Jonathan Lethem, "The Ecstasy of Influence: A Plagiarism," *Harper's* 312, no. 1881 (February 2007): 68.

29. *Mark Twain's Letters,* edited by Albert Bigelow Paine (New York: Harper, 1917), 2: 731.

30. Ralph Waldo Emerson, *Letters and Social Aims,* new and rev. ed., Riverside Edition of *Emerson's Complete Works* (Boston: Houghton Mifflin, 1895), 8: 170.

31. Kenneth Goldsmith, *Uncreative Writing: Managing Language in the Digital Age* (New York: Columbia University Press, 2011), p. 201.

32. Goldsmith, *Uncreative Writing,* p. 215.

33. Keen, "Web 2.0."

34. "Democratic Party (United States)," http://en.wikipedia.org/wiki/Democratic_Party_(United_States), consulted 30 July 2013.

35. "Democratic Party," http://www.conservapedia.com/U.S._Democratic_Party,_history, consulted 30 July 2013.

36. "Wikipedia: Neutral Point of View," http://en.wikipedia.org/wiki/Wikipedia:NPOV, consulted 30 July 2013.

37. "Republican Party," http://www.conservapedia.com/Republican_Party, consulted 11 July 2012.

38. Roger Kimball, *Tenured Radicals: How Politics Has Corrupted Our Higher Education,* 3rd ed. (Chicago: Ivan R. Dee, 2008), p. 250; Dinesh D'Souza, *Illiberal Education: The Politics of Race and Sex on Campus* (New York: Free Press, 1991), p. 229.

39. Randall Stross, "Anonymous Source Is Not the Same as Open Source," *New York Times,* 12 March 2006, http://www.nytimes.com/2006/03/12/business/yourmoney/12digi.html.

40. Howard Rheingold, *Net Smart: How to Thrive Online* (Cambridge, MA: MIT Press, 2012), p. 78.

41. *Truth in Numbers? Everything, According to Wikipedia,* directed by Scott Glosserman and Nic Hill (Glen Echo Entertainment/Underdog Pictures, 2010), http://www.dailymotion.com/video/xqvyck_truth-in-numbers-everything-according-to-wikipedia_tech, consulted 30 July 2013.

42. P. D. Magnus, "On Trusting Wikipedia," *Episteme* (2009): 74–90, at 83–84.

43. *Truth According to Wikipedia.*

44. "Fatally Flawed."

45. Johnny Hendren, "The Art of Wikigroaning," http://www.somethingawful.com/d/news/wikigroaning.php, consulted 30 July 2013.

46. Frank Donoghue, *The Last Professors: The Corporate University and the Fate of the Humanities* (New York: Fordham University Press, 2008), p. 89.

47. Keen, *Cult of the Amateur,* pp. 45, 46.

48. *Truth According to Wikipedia.*

49. Hastings Rashdall, *The Universities of Europe in the Middle Ages* (Oxford: Clarenden, 1895), p. 17.

50. Alexander Blair, "Thoughts on the Advertisement and Diffusion of Knowledge," *Blackwood's Magazine* 16 (1824): 26–33; quoted by Richard Yeo, *Encyclopaedic Visions: Scientific Dictionaries and Enlightenment Culture* (Cambridge: Cambridge University Press, 2001), p. 249.

51. Stephen Turner, "What Is the Problem with Experts?," *Social Studies of Science* 31, no. 1 (2001): 125–49; reprinted in Evan Selinger and Robert P. Crease, eds., *The Philosophy of Expertise* (New York: Columbia University Press, 2006), pp. 159–86, at 165.

52. Andrew Lih, *The Wikipedia Revolution: How a Bunch of Nobodies Created the World's Greatest Encyclopedia* (New York: Hyperion, 2009), p. 191.

53. Lih, *Wikipedia Revolution,* p. 194.

54. *Truth According to Wikipedia.*

55. Robert McHenry, "The Faith-Based Encyclopedia," *Ideas in Action,* 15 November 2004, http://www.ideasinactiontv.com/tcs_daily/2004/11/the-faith-based-encyclopedia.html, consulted 30 July 2013.

56. "Pluto," *Encyclopaedia Britannica,* http://www.britannica.com/EBchecked/topic/465234/Pluto, consulted 30 July 2013.

57. Davis, "Professors' Lie," pp. 7, 15.

58. Keen, "Web 2.0."

59. Jaron Lanier, *You Are Not a Gadget: A Manifesto* (New York: Knopf, 2010), p. 143.

60. Nicholas Carr, *The Shallows: What the Internet Is Doing to Our Brains* (New York: Norton, 2010), p. 139.

61. Carr, *Shallows,* p. 137.

62. Erping Zhu, "Hypermedia Interface Design: The Effects of Number of Links and Granularity of Nodes," *Journal of Educational Multimedia and Hypermedia* 8, no. 3 (1999): 331–58, quoted in Carr, *Shallows,* p. 129.

63. Kevin Kelly, "Scan This Book!" *The New York Times Magazine,* 16 May 2006, quoted by Keen, *Cult of the Amateur,* p. 57.

64. Keen, *Cult of the Amateur,* p. 58.

65. Jürgen Habermas, Acceptance speech for the Bruno Kreisky Prize for the Advancement of Human Rights, 9 March 2006, quoted by Keen, *Cult of the Amateur,* p. 55.

66. Clay Shirky, "A Speculative Post on the Idea of Algorithmic Authority," 15 November 2009, http://www.shirky.com/weblog/2009/11/a-speculative-post-on-the-idea-of-algorithmic-authority/, consulted 30 July 2013.

67. Keen, "Web 2.0."

Chapter Four: Playing the Encyclopedia Game

1. Howard Rheingold, *Net Smart: How to Thrive Online* (Cambridge, MA: MIT Press, 2012), pp. 246–52 and passim.

2. Richard P. Keeling and Richard H. Hersh, *We're Losing Our Minds: Rethinking American Higher Education,* pp. 7, 64, 43.

3. Keeling and Hersh, *We're Losing Our Minds,* p. 43.

4. Gardner Campbell, "More on the Wikipedia Controversy," *Gardner Writes,* http://www.gardnercampbell.net/blog1/?p=296, consulted 31 July 2013.

5. Joan Vinall-Cox, response to "More on the Wikipedia Controversy," *Gardner Writes,* http://www.gardnercampbell.net/blog1/?p=296, consulted 31 July 2013.

6. Cory Doctorow et al., "On *Digital Maoism: The Hazards of the New Online Collectivism,* by Jaron Lanier," *Edge: The Reality Club,* 30 May 2006, http://www.edge.org/discourse/digital_maoism.html, consulted 31 July 2013.

7. Danah Boyd, "Information Access in a Networked World," talk presented to Pearson Publishing, 2 November 2007, http://www.danah.org/papers/talks/Pearson2007.html, consulted 31 July 2013.

8. P. D. Magnus, "On Trusting Wikipedia," *Episteme* (2009): 74–90, at 77–78.

9. Kenneth T. Walsh, "Historians Rank George W. Bush among Worst Presidents: Lincoln and Washington Were Rated as the Best," *U.S. News and World Report,* 17 February 2009, http://www.usnews.com/news/history/articles/2009/02/17/historians-rank-george-w-bush-among-worst-presidents, consulted 31 July 2013.

10. *The Truth According to Wikipedia,* directed by Ijsbrand van Veelen, *Tegenlicht,* 7 April 2008, http://www.youtube.com/watch?v=WMSinyx_Abo.

11. Roger Caillois, *Man, Play, and Games,* translated by Meyer Barash (New York: Free Press, 1961), pp. 9–10.

12. James Titcomb, "First Man to Make 1 million Wikipedia Edits," *The Telegraph,* 20 April 2012, http://www.telegraph.co.uk/technology/wikipedia/9215151/First-man-to-make-1-million-Wikipedia-edits.html, consulted 31 July 2013.

13. Johan Huizinga, *Homo Ludens: A Study of the Play-Element in Culture,* translated by R.F.C. Hull (1949; reprinted Oxford: Routledge, 1998), p. 13.

14. Caillois, *Man, Play, and Games,* pp. 12–13.

15. E. Gabriella Coleman, *Coding Freedom: The Ethics and Aesthetics of Hacking* (Princeton: Princeton University Press, 2013), p. 99.

16. Brian Sutton-Smith, *The Ambiguity of Play* (Cambridge, MA: Harvard University Press, 1997), pp. 9–11.

17. "Wikipedia Community," http://en.wikipedia.org/wiki/Wikipedia_community, consulted 31 July 2013. The reference is to Heng-Li Yang and Cheng-Yu Lai, "Motivations of Wikipedia Content Contributors," *Computers in Human Behavior* 26, no. 6 (2010): 1377–83.

18. Larry Sanger, "The Early History of Nupedia and Wikipedia: A Memoir," http://features.slashdot.org/story/05/04/18/164213/the-early-history-of-nupedia-and-wikipedia-a-memoir, consulted 31 July 2013.

19. Joseph Michael Reagle Jr., *Good Faith Collaboration: The Culture of Wikipedia* (Cambridge, MA: MIT Press, 2010), p. 68.

20. Saabira Chaudhuri, "Editors Won't Let It Be When It Comes to 'the' or 'The'/Wonky Wikipedia Debate: Whether Beatles Article Merits Capital 'T,'" *The Wall Street Journal,* 12 October 2012, http://online.wsj.com/article/SB10000872396390444657804578048534112811590.html#articleTabs%3Darticle, consulted 31 July 2013.

21. "Wikipedia: Edit Warring," http://en.wikipedia.org/wiki/Wikipedia:Revert_war, consulted 31 July 2013; cf. Andrew Lih, *The Wikipedia Revolution: How a Bunch of Nobodies Created the World's Greatest Encyclopedia* (New York: Hyperion, 2009), pp. 127–28.

22. Taha Yasseri et al., "Dynamics of Conflicts in Wikipedia," *Plos One,* http://www.plosone.org/article/info%3Adoi%2F10.1371%2Fjournal.pone.0038869, consulted 31 July 2013.

23. A 2011 poll cited at "Editor's Survey 2011: Executive Summary (http://meta.wikimedia.org/wiki/Editor_Survey_2011/Executive_Summary, consulted 31 July 2013) indicated that over 90 percent of Wikipedia editors are male.

24. "Wikipedia: Lamest Edit Wars," http://en.wikipedia.org/wiki/Wikipedia:Lamest_edit_wars, consulted 8 January 2013.

25. Deacon of Pndapetzim, "How to Win a Revert War," http://en.wikipedia.org/wiki/User:Deacon_of_Pndapetzim/How_to_win_a_revert_war, consulted 31 July 2013.

26. Phoebe Ayers, Charles Matthews, and Ben Yates, *How Wikipedia Works and How You Can Be a Part of It* (San Francisco: No Starch Press, 2008).

27. John Broughton, *Wikipedia: The Missing Manual* (Sebastopol, CA: Pogue Press/O'Reilly, 2008), p. v.

28. "Wikipedia: School and University Projects," http://en.wikipedia.org/wiki/Wikipedia:School_and_university_projects, consulted 31 July 2013.

29. Clay Shirky, *Here Comes Everybody: The Power of Organizing without Organizations* (New York: Penguin, 2008), p. 135.

30. "Wikipedia: Ownership of Articles," http://en.wikipedia.org/wiki/Wikipedia:Ownership_of_articles, consulted 31 July 2013.

31. Ayers, Matthews, and Yates, *How Wikipedia Works,* p. 315.

32. Cullen J. Chandler and Alison S. Gregory, "Sleeping with the Enemy: Wikipedia in the College Classroom," *The History Teacher* 43, no. 2 (February 2010): 247–57, at 255.

33. Chandler and Gregory, "Sleeping with the Enemy," at 254.

34. Chandler and Gregory, "Sleeping with the Enemy," at 255, 254.

35. Robert E. Cummings, *Lazy Virtues: Teaching Writing in the Age of Wikipedia* (Nashville: Vanderbilt University Press, 2009), p. 55.

36. Cummings, *Lazy Virtues,* p. 55.

37. Cummings, *Lazy Virtues,* pp. 16–20; see Yochai Benkler, "Coase's Penguin, or Linux and the Nature of the Firm," *Yale Law Journal* 112, no. 3 (December

2002): 369–447, and *The Wealth of Networks: How Social Production Transforms Markets and Freedom* (New Haven: Yale University Press, 2006), pp. 60–63.

38. Cummings, *Lazy Virtues,* p. 120.

39. Cummings, *Lazy Virtues,* pp. 116, 117.

40. Cummings, *Lazy Virtues,* p. 112.

41. Cummings, *Lazy Virtues,* p. 106.

42. Cummings, *Lazy Virtues,* p. 107.

43. Cummings, *Lazy Virtues,* pp. 120, 121, 122.

44. See Andrea Forte, Vanessa Larco, and Amy Bruckman, "Decentralization in Wikipedia Governance," *Journal of Management Information Systems* 26, no. 1 (Summer 2009): 49–72.

45. Joseph Michael Reagle Jr., *Good Faith Collaboration: The Culture of Wikipedia,* p. 1. See also Broughton, *Wikipedia: The Missing Manual,* pp. 143–229, and Ayers, Matthews, and Yates, *How Wikipedia Works,* pp. 464–66.

46. Kenneth Burke, *Attitudes toward History,* 3rd ed. (Berkeley: University of California Press, 1984), pp. 171, 107.

47. Robert M. Hutchins, "The Great Conversation," *The Great Conversation: A Reader's Guide to Great Books of the Western World* (Chicago: Encyclopaedia Britannica, Inc., 1990), pp. 48–49.

Chapter Five: Tomorrow and Tomorrow and Tomorrow

1. Garrett Hardin, "The Tragedy of the Commons," *Science,* n. s., 162, no. 3859 (13 December 1968): 1243–48, at 1244.

2. Yochai Benkler, *The Wealth of Networks: How Social Production Transforms Markets and Freedom* (New Haven: Yale University Press, 2006), pp. 3, 4, 5.

3. Benkler, *Wealth of Networks,* p. 60.

4. Benkler, *Wealth of Networks,* p. 70.

5. "Wikimedia Movement Strategic Plan Summary," http://wikimediafoundation.org/wiki/Wikimedia_Movement_Strategic_Plan_Summary, consulted 31 July 2013.

6. Joseph Michael Reagle Jr., *Good Faith Collaboration: The Culture of Wikipedia* (Cambridge, MA: MIT Press, 2010), p. 172.

7. Wikipedia: Reliable Sources," http://en.wikipedia.org/wiki/Wikipedia:Reliable_sources, consulted 31 July 2013.

8. "Wikimedia: Terms of Use/Paid Contributions Amendment," http://meta.wikimedia.org/wiki/Terms_of_use/Paid_contributions_amendment, consulted 17 March 2014.

9. "Wikipedia: Flagged Revisions," http://en.wikipedia.org/wiki/Wikipedia:Flagged_revisions, consulted 31 July 2013.

10. "Wikipedia: Flagged Revisions / Reliable Revisions," http://en.wikipedia.org/wiki/Wikipedia:Flagged_revisions/reliable_revisions, consulted 31 July 2013.

11. Mathieu O'Neil, *Cyberchiefs: Autonomy and Authority in Online Tribes* (London: Pluto, 2009), p. 168.

12. Andrew Lih, *The Wikipedia Revolution: How a Bunch of Nobodies Created the World's Greatest Encyclopedia* (New York: Hyperion, 2009), p. 64.

13. See Andrew Keen, *The Cult of the Amateur: How Today's Internet Is Killing Our Culture* (New York: Doubleday/Currency, 2007), pp. 186–89, and Lih, *Wikipedia Revolution,* p. 191.

14. "Animal Diversity Web," animaldiversity.ummz.umich.edu, consulted 31 July 2013.

15. "Internet Encyclopedia of Philosophy," http://www.iep.utm.edu/, consulted 31 July 2013.

16. "Full text of 'Encyclopaedia Biblica: A Critical Dictionary of the Literary, Political, and Religious History, the Archaeology, Geography, and Natural History of the Bible,'" http://archive.org/stream/encyclopaediabib03cheyuoft/encyclopaediabib03cheyuoft_djvu.txt, consulted 20 January 2013.

17. "Jewish Encyclopedia.com," http://www.jewishencyclopedia.com/, consulted 31 July 2013.

18. "The Catholic Encyclopedia," http://www.newadvent.org/cathen/, consulted 31 July 2013.

19. "Encyclopedia of Mormonism," http://eom.byu.edu/, consulted 31 July 2013.

20. "The Encyclopedia of World Problems and Human Potential," http://www.uia.be/encyclopedia, consulted 31 July 2013.

21. "Rocklopedia Fakebandica," http://www.fakebands.com/wiki/index.php?title=Main_Page, consulted 31 July 2013.

22. "Metapedia," http://en.metapedia.org/wiki/Main_Page, consulted 31 July 2013.

23. "Metapedia: Mission Statement," http://en.metapedia.org/wiki/Metapedia:Mission_statement, consulted 31 July 2013.

24. Nicholson Baker, "The Charms of Wikipedia," *The New York Review of Books* 55, no. 4 (20 March 2008): 1–10, at 10.

25. http://deletionpedia.dbatley.com/w/index.php?title=Main_Page, consulted 31 July 2013.

26. Andrew Brown, *A Brief History of Encyclopedias: From Pliny to Wikipedia* (London: Hesperus, 2011), p. 85.

27. Paul Graham, "Startup Ideas We'd Like to Fund," http://ycombinator.com/ideas.html, consulted 31 July 2013.

28. James Gleick, "Wikipedians Leave Cyberspace, Meet in Egypt," *The Wall Street Journal,* 8 August 2008, http://online.wsj.com/article/SB121815517766225 97.html, consulted 31 July 2013.

29. "5 Sterne bei Wiki-Watch"—Nach welchen Kriterien vergeben wir unsere formale Bewertung?," http://blog.wiki-watch.de/?p=80, consulted 31 July 2013.

30. https://addons.mozilla.org/en-US/firefox/addon/wikitrust/, consulted 31 July 2013.

31. "Goals of WikiTrust," http://www.wikitrust.net/goals-of-wikitrust, consulted 31 July 2013.

32. "Anonymous on Fox11," http://www.youtube.com/watch?v=DNO6G4ApJQY, consulted 31 July 2013.

33. James Paul Gee, *What Video Games Have to Teach Us about Learning and Literacy,* rev. and updated ed. (New York: Palgrave Macmillan, 2007).

34. Steven Johnson, *Everything Bad Is Good for You: How Today's Popular Culture Is Actually Making Us Smarter* (New York: Riverhead, 2005), p. 20.

35. Gee, *What Video Games Have to Teach Us*, pp. 221–22.

36. Tim Berners-Lee with Mark Fischetti, *Weaving the Web: The Original Design and Ultimate Destiny of the World Wide Web by Its Inventor* (San Francisco: HarperSanFrancisco, 1999), pp. 157–58.

37. "Wikidata," http://www.wikidata.org/wiki/Wikidata:Main_Page, consulted 31 July 2013.

38. See Lih, *Wikipedia Revolution,* p. 220.

39. David Sano, "Twitter Creator Jack Dorsey Illuminates the Site's Founding Document, Part I," *Los Angeles Times,* 18 February 2009, http://latimesblogs.latimes.com/technology/2009/02/twitter-creator.html, consulted 31 July 2013.

40. "Terms of Use," http://www.snapchat.com/terms/, consulted 4 March 2014.

41. Somini Sengupta, "Facebook's Prospects May Rest on Trove of Data," *New York Times,* 14 May 2012, http://www.nytimes.com/2012/05/15/technology/facebook-needs-to-turn-data-trove-into-investor-gold.html?_r=1&, consulted 31 July.

42. Jacques Derrida, "Structure, Sign, and Play in the Discourse of the Human Sciences," in *Writing and Difference,* translated by Alan Bass (Chicago: University of Chicago Press, 1978), pp. 279–80.

43. Mathew Ingram, "What Wikipedia Can Tell Us about the Future of News," Gigacom, 19 December 2012, http://gigaom.com/2012/12/19/what-wikipedia-can-tell-us-about-the-future-of-news/, consulted 31 July 2013.

44. "About Reddit," http://www.reddit.com/about, consulted 4 March 2014.

45. Hardin, "Tragedy of the Commons," at 1246.

46. Patrik Svensson, "The Landscape of Digital Humanities," *Digital Humanities Quarterly* 4, no.1 (2010), http://digitalhumanities.org/dhq/vol/4/1/000080/000080.html&reason=0#silver2006, paragraph 49, consulted 31 July 2013.

47. Svensson, "Landscape of Digital Humanities," paragraph 14; see Tara McPherson, "Introduction: Media Studies and the Digital Humanities," *Cinema Journal* 48, no. 2 (Winter 2009): 119–23.

48. Svensson, "Landscape of Digital Humanities," paragraphs 53, 86–87.

49. Matthew G. Kirschenbaum, "What Is Digital Humanities and What's It Doing in English Departments?," *ADE Bulletin,* no. 150 (2010): 1–7, at 5, 6; http://mkirschenbaum.files.wordpress.com/2011/01/kirschenbaum_ade150.pdf, consulted 31 July 2013.

50. David Silver, "Introduction: Where Is Internet Studies?," *Critical Cyberculture Studies,* ed. David Silver and Adrienne Massanari (New York: NYU Press, 2006), p. 6.

51. Henry Jenkins, *Convergence Culture: Where Old and New Media Collide* (New York: NYU Press, 2006), p. 266.

52. Joseph Michael Reagle Jr., *Good Faith Collaboration: The Culture of Wikipedia* (Cambridge, MA: MIT Press, 2010), p. 1.

53. Bill Bishop, with Robert G. Cushing, *The Big Sort: Why the Clustering of Like-Minded America Is Tearing Us Apart* (Boston: Houghton Mifflin, 2008), p. 14.

54. Andrew Keen, "Web 2.0: The Second Generation of the Internet Has Arrived. It's Worse Than You Think," *Weekly Standard,* 14 February 2006, http://www.weeklystandard.com/Content/Public/Articles/000/000/006/714fjczq.asp, consulted 31 July 2013.

55. Clay Shirky, "Old Revolutions, Good; New Revolutions, Bad," *Encyclopaedia Britannica Blog,* 14 June 2007, http://www.britannica.com/blogs/2007/06/old-revolutions-good-new-revolutions-bad/, consulted 31 July 2013.

56. O'Neil, *Cyberchiefs,* p. 175.

57. Lih, *Wikipedia Revolution,* p. xvii.

58. *Truth in Numbers? Everything, According to Wikipedia,* directed by Scott Glosserman and Nic Hill (Glen Echo Entertainment/Underdog Pictures, 2010), http://www.dailymotion.com/video/xqvyck_truth-in-numbers-everything-according-to-wikipedia_tech, consulted 31 July 2013.

59. Benkler, *Wealth of Networks,* p. 293.

60. Rheingold, *Net Smart,* p. 182.

INDEX